Green Chemistry

Authors

Dr. Rajni Kant Sharma

Dr. Sushila Singh

Dr. Monika Moond

Anuradha Beniwal

Jyoti Rani

Rajesh Moond

Integrated Publications ™
New Delhi

Published By: *Integrated Publications* TM

Integrated Publications
H. No. - 3 Pocket - H34, Sector - 3,
Rohini, Delhi-110085, India
Email – info@integratedpublications.in

Authors: *Dr. Rajni Kant Sharma, Dr. Sushila Singh, Dr. Monika Moond, Anuradha Beniwal, Jyoti Rani and Rajesh Moond*

Edition: *1st*

Publication Year: *2023*

Pages: *99*

ISBN: *978-93-5834-007-5*

Book DOI: *https://doi.org/10.22271/int.book.333*

Price: *₹515/-*

Contents

S. No	Chapters	Page No.
1.	Principles of Green Chemistry	01-16
2.	Alternative Energy Sources	17-40
3.	Catalysts and Reagents	41-61
4.	Inherent Safety	62-77
5.	Future Trends in Green Chemistry	78-99

Chapter - 1

Principles of Green Chemistry

Abstract

The development of chemical procedures and goods that restrict the usage and manufacturing of dangerous substances is referred to as "green chemistry". Green chemistry seeks to safeguard the environment, and there are numerous ways to do this.

At the molecular level, "green chemistry," a subject that is still in its infancy of growth, strives for sustainability. Because it has the potential to leverage chemical innovation to simultaneously achieve economic and environmental goals, the subject has received a lot of interest in the last ten years. In this chapter, the Twelve Principles that form the foundation of Green Chemistry have been thoroughly examined.

The green chemistry techniques that have been developed so far encompass the majority of the fields of chemistry, including biochemistry, technological, toxicological, organic, physical, polymer, environmental, inorganic, etc. The dual objectives of environmental safety and financial gain can be attained by utilizing several popular green processes, such as biocatalysis, catalysis reaction media (supercritical fluids, ionic liquids, and water), the use of alternative renewable feedstock (biomass), new synthetic pathways (photocatalytic reaction), and reaction conditions (microwave irradiation). This chapter gives instances of recent developments that show how Green Chemistry reduces the environmental impact of chemical processes and technology.

Keywords: green chemistry ionic liquids, biomass, supercritical fluids.

Introduction

The idea of "green chemistry" is entrenched in twelve guidelines and principles intended to minimize or eliminate harmful components from the synthesis, manufacture, and usage of chemical products. Therefore, it is advisable to reduce or stop using chemical materials that are unsafe for both the environment and human health [1]. The terminology and the idea of green chemistry were initially created around the beginning of the 1990s, or more than 30 years ago. The earliest main programs were headquartered in Italy, the United States, and the United Kingdom. Since then, green chemistry has acquired wide spread support on a global scale, prompting the formation of literally hundreds of programs and governmental initiatives [2].

When R&D project first begins, the design scientist uses these concepts as a set of tools to help them foresee potential problems down the road. Given that the majority of material scientists lack the formal education required to handle problems connected to regulatory implications, human health, and the environment, the twelve principles act as a guide for creating products and processes that would be less harmful to the environment while preserving or improving the economic cost and product performance. Industrial and engineering chemistry now places a high priority on minimizing environmental impact when designing processes, and this trend is expected to continue for the next 100 years [3]. New photocatalytic reaction, alternative reaction conditions (Mechanochemistry, ultrasound, and microwave activation), bio-catalysis, alternate reaction media (ionic liquids, water, supercritical fluids), catalysis, as well as the usage of alternative viable raw matter (biomass), are some of the dominant ways that green chemistry achieves its objectives of environmental protection and financial gain [4].

This chapter attempts to advance knowledge of the elementary concepts and procedures of green chemistry and to illustrate how green chemistry lessens the natural world impact of technologies and chemical reactions through examples and prevailing paradigms [5,6].

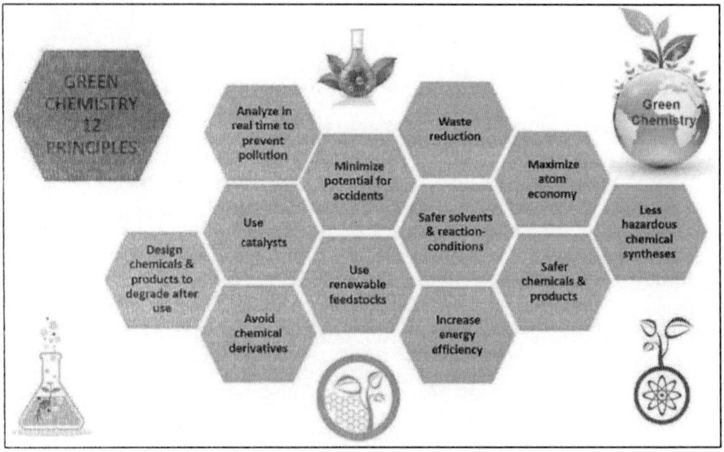

Fig 1: Green chemistry principles

Green Chemistry

One of the best subfields of chemistry is "green chemistry," which takes an ecological position focused on minimizing or eliminating the usage of hazardous materials in chemical reactions as well as minimizing hazardous and lethal byproducts and intermediates. "Green chemistry" is defined as the "design, development, and application of chemical products and processes that reduce the use and generation of substances that are hazardous to human health and the environment". Green chemistry is the creation of ecologically friendly chemicals and practices, according to the United States Environmental Protection Agency (US EPA). Sustainable chemistry is another name for green chemistry. By minimizing or reducing the dangers of chemical reagents, solvents, products, and feedstock, it reduces pollution at its source [7].

Principles of Green Chemistry

The twelve concepts of green chemistry were developed by Paul T. Anastas and John C. Warner of the US EPA in 1998. These guidelines address a broad range of synthetic organic synthesis, including the creation of organic synthesis procedures to lower waste and by-product generation, employ less hazardous chemicals and raw materials, and utilize more safe solvents, biocatalysts, and renewable raw materials [8].

1. Prevention

Instead of treating and cleaning waste that has already been produced through a chemical process, it is preferable to prevent it. Therefore, it is best to do a synthesis by adhering to a pathway where the waste formation is either minimal or non-existent. For example,

A green manufacturing process of sertraline (Scheme 1) in which the addition of ethanol makes smaller the need for the use, distilling, and recovering solvents such as hexane, toluene, dichloromethane, and tetrahydrofuran, [9].

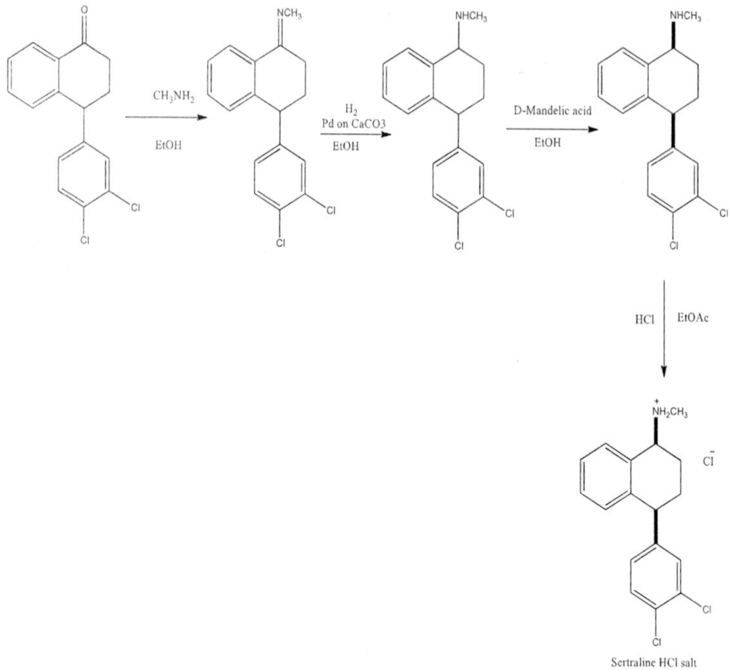

Scheme 1: Green synthesis of Sertraline

2. Atom economy

Instead of producing undesirable by-products or waste, synthetic methods ought to be created to maximize the inclusion of all raw materials used in the process in the final product. Berry Trost of Stanford University created the idea of an atom economy in 1991. The atom economy is the percentage difference between the total number of reactants and the desired product's relative molecular masses.

% atom economy = (Molecular weight of desired product/Molecular weights of all reactants) * 100

To demonstrate the benefits of atom economy, consider the synthesis of ibuprofen. Ibuprofen is synthesized in six steps by the Boots method, compared to just three steps by the Hoechst method. The yield from boots synthesis was 40% while the Hoechst process gives a yield of 77%. One of the "green synthesis" methods is this one [10, 11]. Scheme 2 compares the "green" Hoechst ibuprofen synthesis with the traditional Boots synthesis.

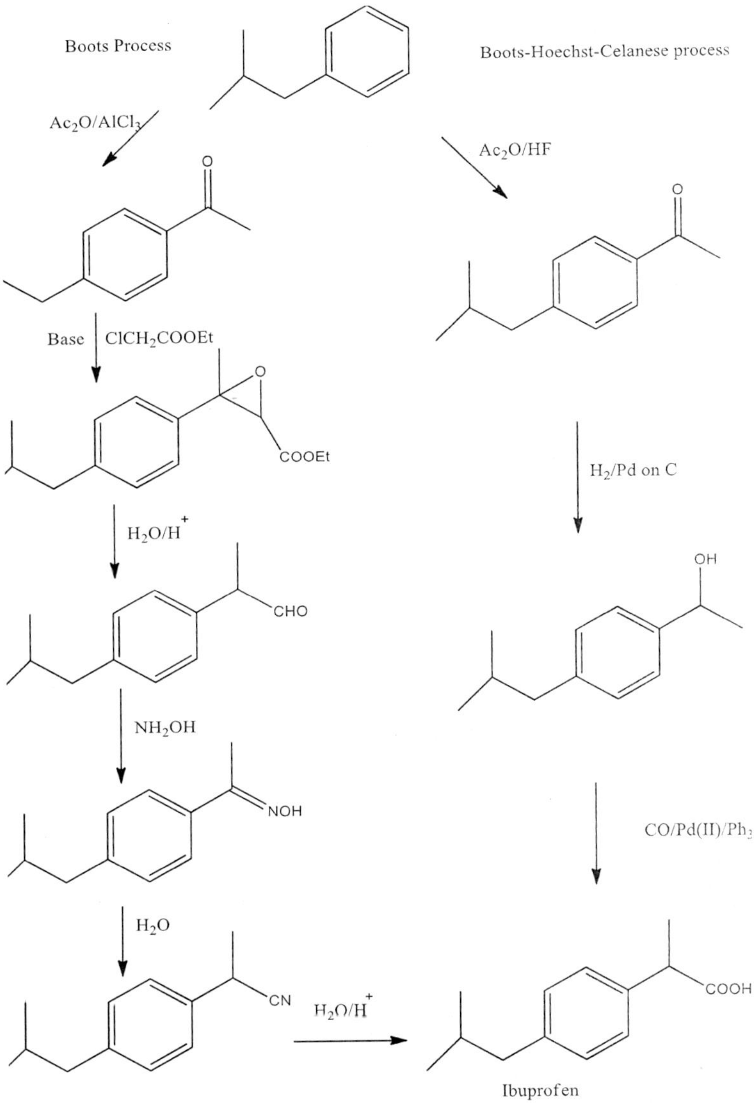

Scheme 2: Comparison of Boots and the "Green" Process of Hoechst for the synthesis of Ibuprofen

The procedure of evaluating the ratios of waste matter and the applicable outcome is known as the "E-factor" in the atom economy. The ratio of the mass of waste (kg) per kilogram of product is used to calculate the E-factor. [12].

E-factor = kg of waste/kg of product

The chunk of waste produced decreases as the value of the E factor approaches zero (factor E ~ 0), making the process more environment friendly and sustainable. Calculating the Environmental Quotient (EQ), a well-liked Green Chemistry Index, requires knowledge of the E-factor's value. It is a combination of the Q value and the E-Factor. The Q value, also known as the environmental risk quotient, has a relationship with the ecotoxicity of the waste generated during an industrial procedure or organic process. In contrast to heavy metals and their salts, which have a Q value between 100 and 1000, sodium chloride has a Q value of 1 [13].

Pharmaceutical products, for example, can have an E-factor of 100, which means that for every kg of product produced, 100 kg of waste is also produced. This is necessary for high-purity products.

3. Use of less hazardous chemicals

Synthetic methods should be created wherever possible to use and produce chemicals that are safe for both the environment and people. For instance, the synthesis of polycarbonate (Scheme 3) relies on the elimination of the solvent dichloromethane (CH_2Cl_2) and the poisonous carbonyl dichloride ($COCl_2$) [14].

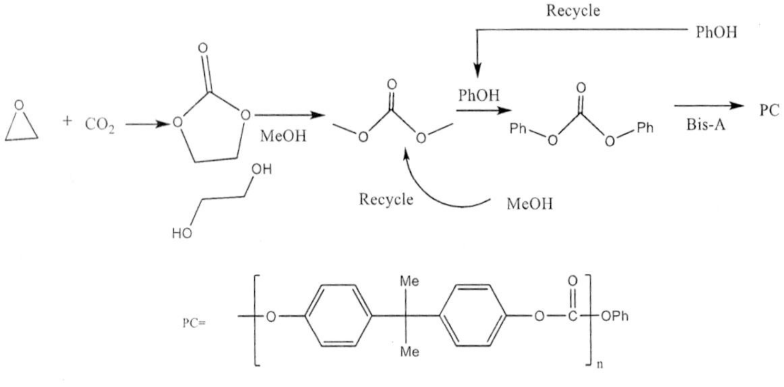

Scheme 3: Synthesis of polycarbamate (PC)

4. Designing Safer Chemicals

Chemical goods should be made in a way that not only fulfills their intended purpose but is also less harmful over time. This concept is used to create novel insecticides and pesticides that are highly specific to the target species. The invention of an oxidant activator for hydrogen peroxide, which allows hydrogen peroxide to replace chlorine bleach in paper manufacture, is the most prevalent example. Another example is the development of polymers of polyphenylene sulfone (PPSU), a novel plastic that stands out for having a unique blend of advantageous mechanical, flame-resistant, and environmental properties [15].

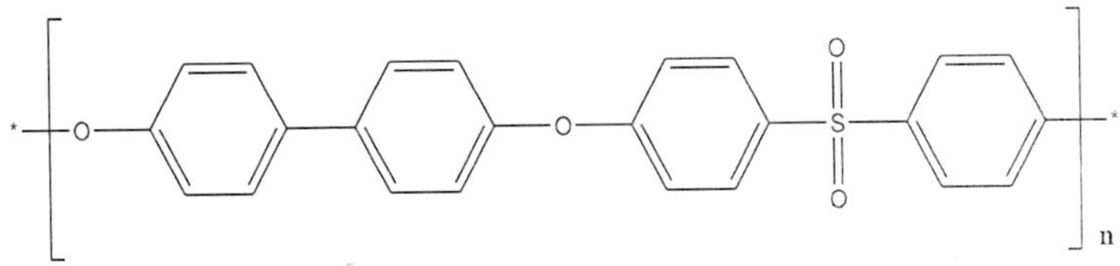

Polyphenylsulfone : PPSU

Schene 3: PPSU

5. Safer Solvents and Auxiliaries

When feasible, it is best to avoid using auxiliary substances like solvents, separating agents, etc. If its use is unavoidable, it should be done as gently or harmlessly as possible. Although it aids in the synthesis of a material, an auxiliary ingredient does not mix in to form the chemical. Scheme 4 depicts an example reaction for stereoselective halogenation [16].

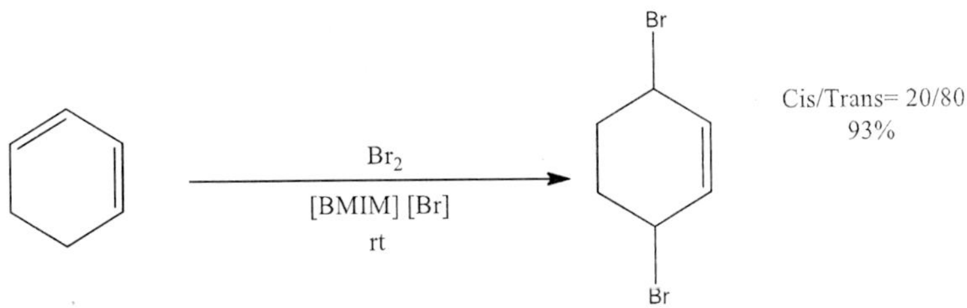

BMIM = 1-butyl-3-methylimidazolium

Scheme 4: Representative reaction for stereoselective halogenation

The selection of suitable substitutes for organic solvents should consider process safety, worker safety, process sustainability, and environment safety, in accordance with the principles of green chemistry. The solvent should be straight forward to use, low volatility, and chemically and physically stable. These are referred to as green solvents. Examples include ionic liquid water and liquid carbon dioxide (supercritical CO_2). Diels-Alder reaction with and without ionic liquid is shown in schemes 5 and 6 respectively.

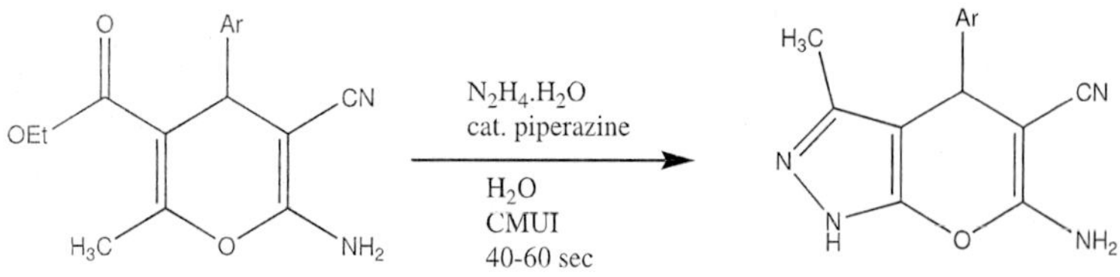

Scheme 5: Diels-Alder reaction

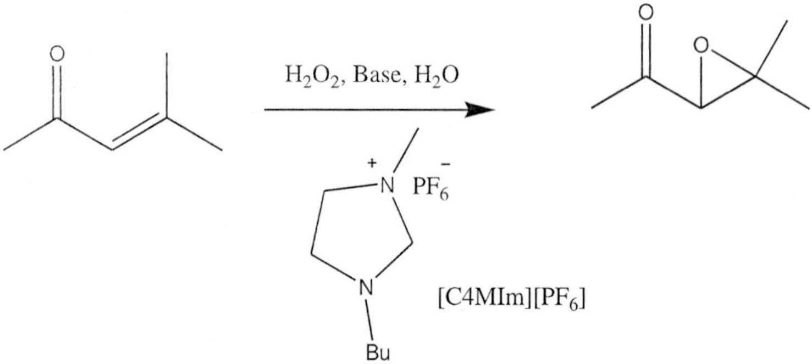

Scheme 6: Diels-Alder reaction in ionic liquid

Ionic liquids, which do not vaporise quickly and have a lower vapor pressure than volatile organic substances, enable safer chemical reactions. Hydrogenation of alkene in $scCO_2$ solvent is shown in scheme 7 [17].

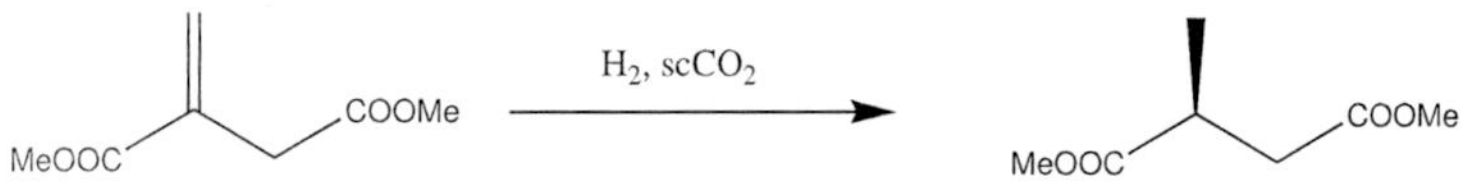

Scheme 7: Hydrogenation of alkene in $scCO_2$ solvent

6. Design for Energy Efficiency

Energy requirements for chemical processes should be kept to a minimum in order to minimize their negative effects on the environment and economy. All reactions should, if at all feasible, be carried out at low pressures and temperatures. For instance, the transformation of benzyl chloride into benzyl cyanide is completed in a very short period of time if we use a phase transfer catalyst (Scheme 8).

$$C_6H_5CH_2Cl + aqKCN \longrightarrow C_6H_5CH_2CN + KCl$$

Benzyl Chloride — Benzyl Cyanide

Scheme 8: Synthesis of Benzyl Cyanide

Conventionally, the reaction has been carried by heating on wire gauze, in an oil bath or heating mantles. However, it is now conceivable to use microwaves, sonication, or photochemistry to provide a reaction with energy (Scheme 9). Examples are

(a) $$C_6H_5CONHC_6H_5 \xrightarrow[\text{MW, 12Min}]{} C_6H_5COOH$$

(b) $$RCOOH + ROH \xrightarrow[H_2SO_4\text{, Ultrasound}]{\text{Esterification}} RCOOR$$

Scheme 9: Reactions using (a) microwave (b) sonication

In a greenhouse where tomatoes are grown, waste steam from a nearby chemical company used to make ammonia is used to heat the space. Since greenhouse CO_2 concentrations are below 50%, plant development is possible. Additionally, carbon dioxide can be employed as a bi-activator for tomato seed growth by being extruded from greenhouse gas [18].

7. Use of Renewable Feedstocks

Raw materials or feedstock should be natural whenever it is technically and financially feasible to do so. Continuous usage of non-renewable resources can lead to their depletion. Therefore, these resources are not thought of as sustainable. Utilizing renewable raw materials is more practical than using various plastics. One of the best examples is to use of biodegradable packaging which has global demand in the food industry. The principle also assumes the use of renewable energy sources like solar, biomass, wind, hydro, and biofuels. Energy use and CO_2 emissions are reduced when bioplastics are made using renewable raw materials. Synthesis of THF shown in schemes 10 and 11.

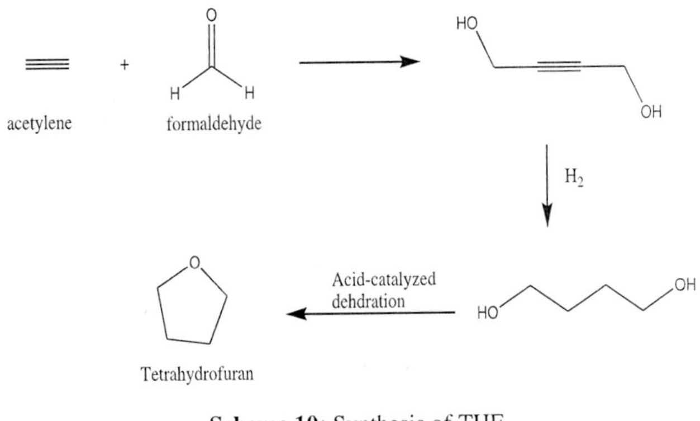

Scheme 10: Synthesis of THF

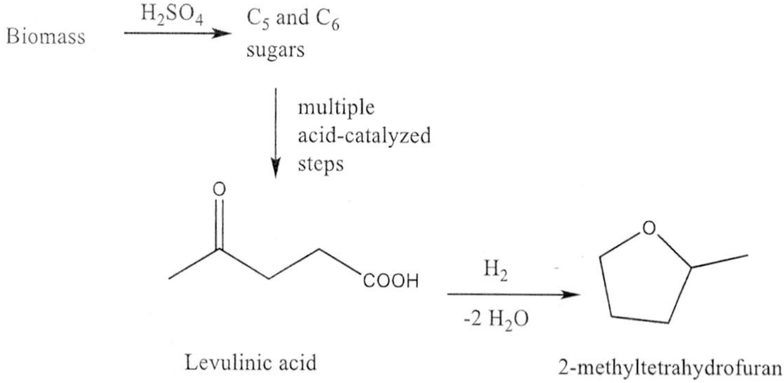

Scheme 11: Green Synthesis of THF

8. Reduction of Derivatives

Derivatization steps that are unnecessary might produce waste and should be minimised or avoided if possible. Derivatization is defined as the application of blocking groups, protection/deprotection, and short-term alteration of chemical/ physical procedures often used during chemical synthesis. Such derivatization requires additional reagents and can produce waste. Using enzymes rather than protecting groups allows for the industrial synthesis of semi-synthetic antibiotics like ampicillin and amoxicillin [19]. Synthesis of semi-synthetic penicillins shown in Scheme 12.

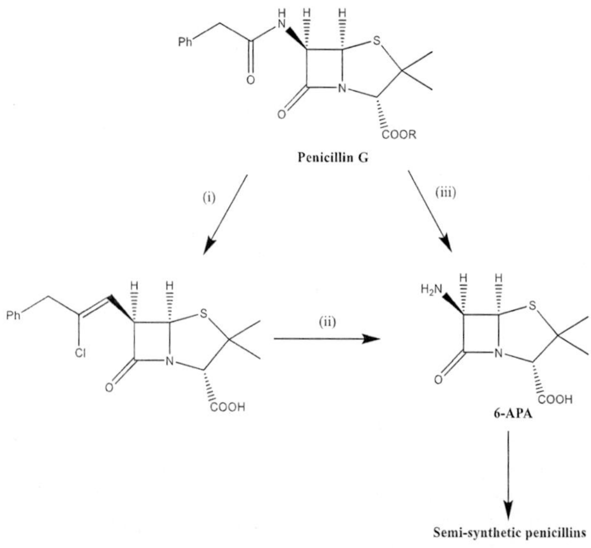

Scheme 12: Synthesis of semi-synthetic penicillins

9. Catalysis

Catalytic reagents that are as selective as feasible outperform stoichiometric reagents in performance. The catalysis principle uplifts the usage of biodegradable catalysts, which require limited amount of energy and do not contain organochlorine compounds, as well as conserving water and generating less wastewater, in order to protect the environment. In classical catechol synthesis, undesirable by-products are formed when are derived from benzene (non-renewable feedstock) while D-glucose (renewable raw materials) in the incidence of genetically adapted *Escherichia coli* produces no by-products [20]. Comparison of classical and biocatalytic-processes for Catechol synthesis shown in Scheme 13.

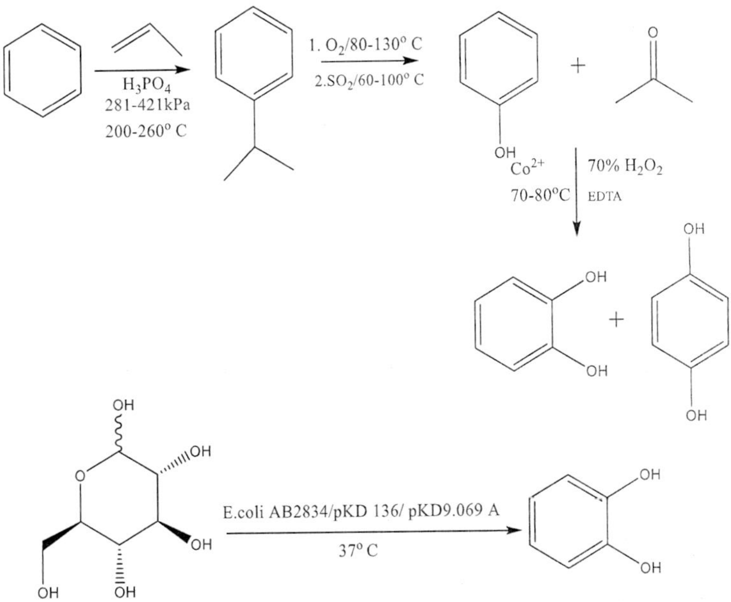

Scheme 13: Comparison of classical and biocatalytic processes for Catechol synthesis

10. Design for Degradation

Chemical products should be made in such a way that, after serving their purpose, they break down into safe degradation products and are no longer remain in the environment. Recycling serves the dual purpose of preventing the growth of pollutants and reducing the amount of waste that is produced [21]. For instance, inadequate degradation led to an accumulation of tetra propylene-alkyl benzenesulfonate (TPPS), a surfactant used in laundry detergents. When the methyl-branched chain is swapped out for a linear carbon chain, TPPS is less persistent. A general example is the substitution of TPPS for linear alkylbenzene sulfonate (LAS) [22]. Figure 4 shows various surfactants design.

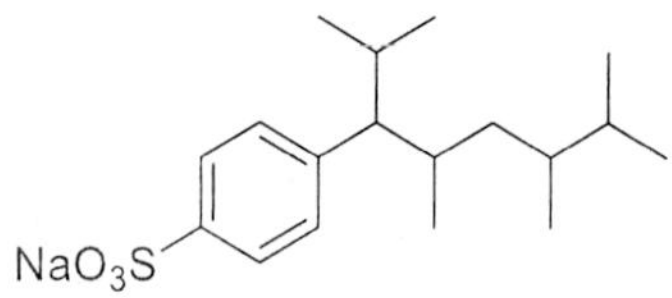

tetrapropylene alkyl benzene sulfonate(TPPS)

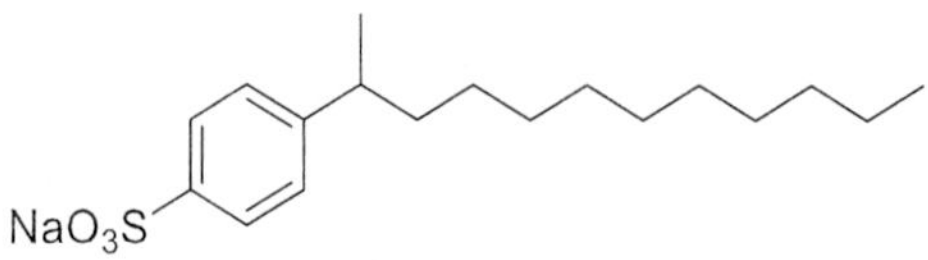

Linear alkyl benzene sulfonate(LAS)

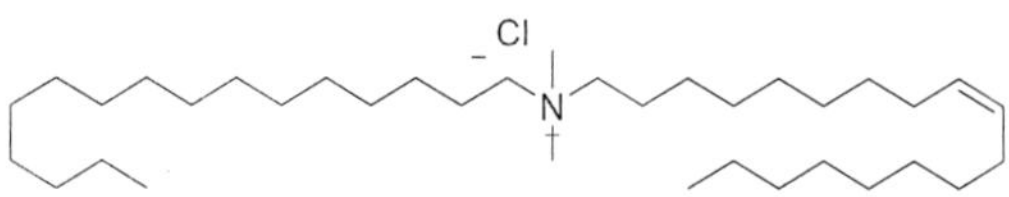

Di(hydrogenated)tallowdimethylammonium chloride (DHTDMAC)-Biodegradability test: 0.5%

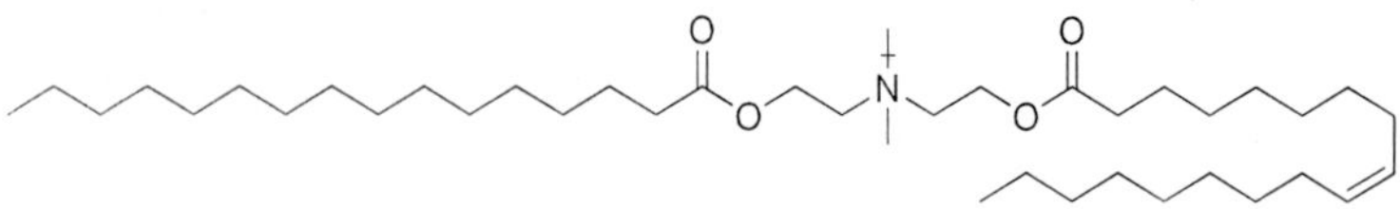

Di(ethyl ester)dimethylammonium chloride (DEEDMAC)- Biodegradibilty test: 76%

Scheme 14: Integrate biodegradability into surfactant design

11. Real-Time Analysis for Pollution Prevention

To enable real-time monitoring and management of the process before the creation of dangerous chemicals, the analytical techniques must be upgraded. This concept enables real-time monitoring of the chemical production process to stop the creation of dangerous compounds. Errors can be avoided by closely watching the production process at every level.

12. Inherently Safer Chemistry for Accident Prevention

To reduce the likelihood of chemical accidents including fires, the release and storage of hazardous chemicals, and explosions, care should be taken while selecting the chemicals and substance forms utilized in the chemical process. For instance, the use of supercritical CO_2 has increased in place of organic solvents because it is non-toxic, non-explosive, and environmentally acceptable. In the development of chemicals and processes, various hazards must be considered, including physical hazards for example explosion, toxicity, or flammability, and global risks to prevent accidents similar to the Bhopal tragedy [23].

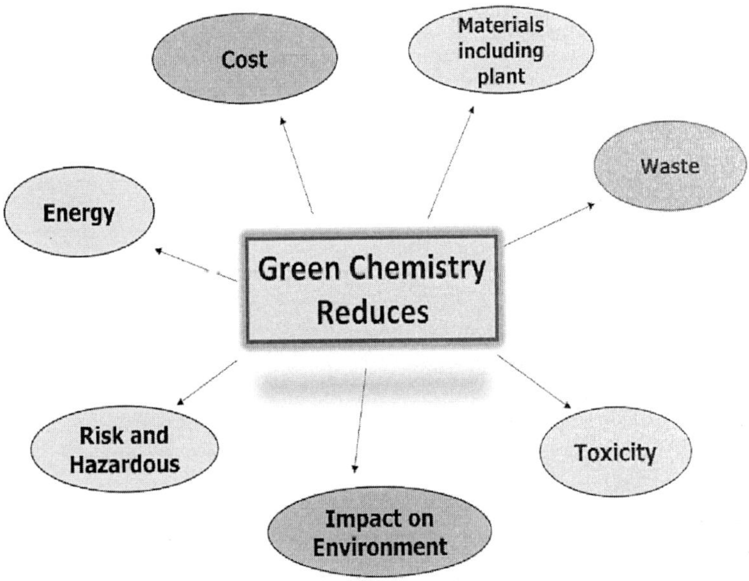

Fig 2: Green Chemistry Reduces

Conclusion

The philosophy of "green chemistry" is crucial for the twenty-first century and beyond. Due to societal and governmental demands regarding environment issues, the chemical and related sectors have taken this attitude very seriously. In addition, industries have been obliged to consider biotechnology and green alternatives in order to achieve efficient manufacturing processes due to the depletion of fossil fuels and rising worldwide competitiveness. It can be very challenging to assess how environmentally friendly a process is. A number of indicators are being developed right now to gauge the process features.

The green chemistry hierarchy is divided into tiers, each of which is focused on a different set of indicators. It is probably hard to establish a connection between the procedure and the environmental impact on a broader scale or to put a price on this impact.

Comparing similar procedures is simple, but contrasting various technology is highly challenging. Additionally, there hasn't been enough study done on the physical and chemical information, the short- and long-term effects of chemicals on the environment, or the interactions between people and the ecosystem. People sometimes realise their errors in hindsight, but by then it can already be too late.

Along with the ideal process and product, one must also consider the ideal user. By selecting, utilising, and disposing of products responsibly, the user can support the green chemistry project. In order to support sustainable development, the user may be able to exert pressure on manufacturing companies to alter their practices. Five times as many jobs could be created through product recycling as through remediation. The first two of the four Rs-Reduce, Recycle, Reuse, and Remediate are associated with green chemistry. The third R describes a responsible user, and the fourth R should only be used as a last resort. This book discusses the most recent advancements in green chemistry and green process technologies using relevant practical examples.

References

1. Abdussalam-Mohammed, W., Ali, A. Q., & Errayes, A. O. (2020). Green chemistry: principles, applications, and disadvantages. *Chem. Methodol*, *4*(4), 408-423.

2. Anastas, P., & Eghbali, N. (2010). Green chemistry: principles and practice. *Chemical Society Reviews*, *39*(1), 301-312.

3. Jessop, P. G., Trakhtenberg, S., & Warner, J. (2009). The twelve principles of green chemistry.

4. Ivanković, A., Dronjić, A., Bevanda, A. M., & Talić, S. (2017). Review of 12 principles of green chemistry in practice. *International Journal of Sustainable and Green Energy*, *6*(3), 39-48.

5. Jukić, M., Djaković, S., Filipović-Kovačević, Ž., Kovač, V. and Vorkapić-Furač, J. (2005): Dominant trends of green chemistry. Kem Ind 54 (5): 255-272, In Croatian.

6. Margetić, D. (2005): Mechanic-chemical organic reactions without the use of solvents. Kem Ind 54 (7-8): 351-358, In Croatian.

7. Ivanković, Anita, Ana Dronjić, Anita MartinovićBevanda, and StanislavaTalić. "Review of 12 principles of green chemistry in practice." International Journal of Sustainable and Green Energy 6, no. 3 (2017): 39-48.

8. Valavanidis, A., ThVlachogianni, and K. Fiotakis. "Laboratory experiments of organic synthesis and decomposition of hazardous environmental chemicals following green chemistry principles." In International Conference "Green Chemistry and Sustainable development", Thessaloniki, pp. 25-26. 2009.

9. Sheldon, Roger A. "Utilisation of biomass for sustainable fuels and chemicals: Molecules, methods and metrics." Catalysis Today 167, no. 1 (2011): 3-13.

10. Anastas, Paul T., Robert Boethling, and AdelinaVoutchkova-Kostal. Green processes: designing safer chemicals. John Wiley & Sons, 2014.

11. Anastas, Paul T., Mary M. Kirchhoff, and Tracy C. Williamson. "Catalysis as a foundational pillar of green chemistry." Applied Catalysis A: General 221, no. 1-2 (2001): 3-13.

12. Zhang, Yi, Bhavik R. Bakshi, and E. SahleDemessie. "Life cycle assessment of an ionic liquid versus molecular solvents and their applications." Environmental science & technology 42, no. 5 (2008): 1724-1730.

13. Heinzle, Elmar, Diane Weirich, Franz Brogli, Volker H. Hoffmann, GuntramKoller, Marcel A. Verduyn, and KonradHungerbühler. "Ecological and economic objective functions for screening in integrated development of fine chemical processes. 1. Flexible and expandable framework using indices." Industrial & engineering chemistry research 37, no. 8 (1998): 3395-3407.

14. Welton, Tom. "Solvents and sustainable chemistry." Proceedings of the Royal Society A: Mathematical, Physical and Engineering Sciences 471, no. 2183 (2015): 20150502.

15. Hill Jr, H. Wayne, and D. G. Brady. "Properties, environmental stability, and molding characteristics of polyphenylenesulfide." Polymer Engineering & Science 16, no. 12 (1976): 831-835.

16. Pavlinac, Jasminka, Marko Zupan, Kenneth K. Laali, and StojanStavber. "Halogenation of organic compounds in ionic liquids." Tetrahedron (Oxford. Print) 65, no. 29-30 (2009).

17. Kärkkäinen, Johanna. Preparation and characterization of some ionic liquids and their use in the dimerization reaction of 2-methylpropene. Vol. 84. Oulu: University of Oulu, 2007.

18. Garnett, Tara. "Fruit and vegetables & UK greenhouse gas emissions: exploring the relationship." UK: Food and Climate Research Network, University of Surrey (2006).

19. Jiménez-González, Concepción, Peter Poechlauer, Quirinus B. Broxterman, Bing-Shiou Yang, David Am Ende, James Baird, Carl Bertsch et al. "Key green engineering research areas for sustainable manufacturing: a perspective from pharmaceutical and fine chemicals manufacturers." Organic Process Research & Development 15, no. 4 (2011): 900-911.

20. Draths, K. M. "JW Frost in Green Chemistry, Frontiers in Benign Chemical Syntheses and Processes." (1998): 150-177.

21. Williams, R. T. "Human health pharmaceuticals in the environment—an introduction." Human pharmaceuticals: assessing the impacts on aquatic ecosystems (2005): 1-46.

22. Anastas, Paul, and Nicolas Eghbali. "Green chemistry: principles and practice." Chemical Society Reviews 39, no. 1 (2010): 301-312.

23. Stellman, J. M. "Encyclopaedia of occupational health and safety. V. 1 The body, health care, management and policy. V. 2 Hazards. V. 3 Industries and occupations. V. 4 Guides, indexes, directory. 4." (1998).

Chapter - 2

Alternative Energy Sources

Abstract

The renewable energy sources (RES) has applications in most cases (wind, solar, biomass, hydropower plant) and these are considered as an alternative source of energy and don't produce harmful pollutants and don't increase foreign matter. In the coming decades, the applications of RES increase continuously and it has the potential to generate electricity. The present study included how RES is beneficial for the environment and over non-renewable energy sources. The green efforts work to raise awareness of green chemistry as an alternative to current techniques of organic synthesis and offer new tools, knowledge, and designs for organic synthesis that will help society's economy to protect the environment and public health.

Keywords: Renewable energy source, environment, green chemistry, synthesis

Introduction

Alternative energy sources are deemed "clean" since they don't release hazardous pollutants and do not increase their carbon footprint. It includes all renewable energy sources. In the coming decades, these forms of energy will replace the usage of fossil fuels and coal as the predominant form of energy production. The phrase "alternative energy sources" officially refers to energy sources that are more environment friendly and will significantly improve the environment. Alternative energy technology has the main advantages of not running out and not requiring the same cost and destructive extraction methods. Various forms of renewable energy are those found in nature. Renewable refers to something that "fills up naturally and can never run out." Common energy sources include coal, gasoline, and natural gas. These have been researched and used for many centuries in power plants to generate electricity; they are occasionally referred as fossil fuels [1]. Renewable energy sources are commonly viewed as emissions free, environment friendly energy sources [2]. The first sign of the industrial revolution was an increase in greenhouse gases in the atmosphere. Emissions of greenhouse gases are to blame for the rise in global temperature over the previous century [3]. The primary causes of global warming are the methods used to produce electricity using fossil fuels, which release greenhouse gases like carbon dioxide (CO_2) into the atmosphere. Coal-based power plants were responsible for the production of almost 40% of all greenhouse gases [4]. The atmosphere is heated by these gases because they absorb heat and store it there. We must acknowledge that our earth will be less liveable if the temperature of the earth will rise likewise. The increased temperature causes glacial ice to melt, which raises the sea level. It will be disastrous as many different islands will submerge. The electricity sector must quickly decarbonize in order to address these issues and meet the 2°C warming target outlined in the Paris Agreements [5]. Furthermore, alternative energy sources are being used more frequently. Power plants using natural gas produce 50% less CO_2 than those using other fossil fuels. Therefore, natural gas could be used as a more sustainable source of energy until renewable energy sources are more widely available [6]. Access to reasonably priced renewable energy is the seventh goal of the UN's 17 sustainable development goals for a greener world. Renewable energy sources (RES) are thought to be emission-free and easily available in nature [7]. Combined geothermal, concentrated solar thermal, oceanic, and bio-power sources account for just over 6% of the remaining RES. Ocean power's contribution is only 0.02 percent, but it has more potential to increase. In several large nations including US, the ratios of Mexico, China, Australia, Russia, and other countries range from 20 to 40%. Many nations are devoted to reducing their reliance on fossil fuels by adopting RES for preserving the environment. The UK's overall demand can be satisfied by only about 28% of tidal energy [8]. According to life cycle research, approximately 70–80% of the negative effects on the environment is because of materials used in the production of wind turbines. As a result, procedures are created to choose the optimal location for power plants to be built in order to minimize adverse environmental effects [9]. Standards for economic, technological, and environmental considerations are laid out. Only photovoltaic (PV) modules and wind farms are the subjects of the investigation. Additionally, there is a brief discussion on the wave and tidal power plants. Multiple electrochemical impacts on soil are discussed which are caused by sewage of wind, solar, and biomass power plants [10]. The report outlines recent innovations to reduce these worries. However, it falls short of emphasising other RES power plants with a base. Renewable energy sources can have an influence on the environment even though they are frequently seen as essentially emission-free for example, solar [11], wind [12] and hydropower development [13] to create these technologies with sustainable development in mind practices [14, 15]. Despite its enormous potential and rising popularity, solar energy is still merely a small percentage of the world's energy supply [16]. The ongoing effort to reduce carbon emissions, which has been a significant global issue in the contemporary social, economic, and environmental context, represents yet another enormous opportunity for solar research [17]. The expansion of renewable energy (RE) is mostly a process driven by a policy with the goals of enhancing energy security, defending the environment, and promoting economic prosperity [18]. RES provides numerous environmental and social advantages, including local job and welfare creation [19], money creation, reduction of rural residents' ageing population [20], diversification of rural economies, and increased social cohesion [21]. By using endogenous resources reduction of global warming and environmental improvement can occur through sustainable energy technology [22]. Focus is mostly placed on "greening" these three elements when attempting to make a chemical reaction "green". By "greening," we mean employing safe solvents or conducting reactions without the use of solvents, using other, more potent reagents and catalysts, and adjusting the reaction conditions through the use of economical, environment friendly techniques. This chapter discusses alternative energy sources as well as newer techniques for kinetic activation of molecules in chemical reactions. Several green approaches have been investigated and discussed in order to decrease the energy requirements of reactions, enhance the speed of the reaction, increase product yield, and eliminate unwanted side products.

Types of Energy Sources

1. Renewable energy sources

Renewable energy sources are flow-limited but naturally replenishing. Although they have an almost endless lifespan, they have a limited amount of energy available per minute. Solar, wind, biomass, ocean waves, hydroelectricity and geothermal heat are some of these sources [23].

2. Non-renewable energy sources

Natural resources that cannot be renewed quickly enough to meet demand are known as non-renewable energy sources. Gas, coal, and oil are the three main categories of fossil fuels used as non-renewable energy sources. While there are many domestic uses for fossil fuels, about 70% of the energy used in the industrial processes comes from non-renewable sources.

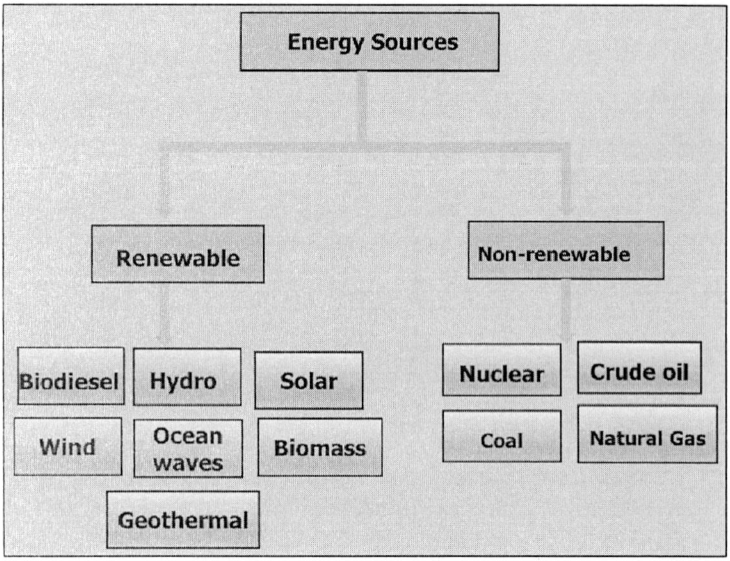

Fig 1: Types of energy sources

Renewable Energy Sources

1. Solar energy

Given their constantly improving output efficiency and versatility, solar technologies are among the best promising renewable energy sources. Solar energy is advantageous due to its inherent properties, especially for developing nations that are situated in areas with easy access to the sun's rays. Through the use of photovoltaic cells, solar energy is captured straight from the Sun and transformed into electrical energy. Solar thermal power is the use of heat from the sun to produce electricity under specific conditions.

Solar energy can be converted using a variety of power technologies. Some of these are as follows:

- Indirect solar energy is produced by hydroelectric power plants.
- When sunlight strikes a light mill, it rotates the vanes.
- A thermal collector receives concentrated sunlight, which heats up the surface. A fluid carries the heat away.
- Direct electricity production from solar energy occurs in photovoltaic cells.
- The only possible disadvantage of solar panels is that they are susce

ptible to degradation over time and are not completely weatherproof in countries with variable weather.

- Sunlight is directed onto a reflecting pipe located externally, which allows it to enter building interiors and provide additional lighting [24].

Solar Power Plants

Solar energy has long been recognised as clean energy. It does not emit carbon dioxide (CO_2) [25]. Nearly 140PW of solar energy reach the surface of the global, out of which only 36PW are put to use [26]. Photovoltaic technology (PVT) and concentrating solar power (CSP) are the best impressive methods for obtaining energy from the sun and the following subsections listed below demonstrate how PVT and CSP affect the ecosystem.

Photovoltaic technology

In recent years, the use of photovoltaic devices, which use semiconductors to directly convert solar energy into electrical energy, has grown [27]. Increased photovoltaic technology efficiency has resulted from energy scientists' tireless study into the potential of solar power. A promising achievement has been made in the hybrid solar cell such as ((CH_3NH_3) PbI_3), a claimed 18% efficiency increase. Commercial thin-film cells made up of telluride, Cadmium, Copper, Indium, amorphous silicon, and gallium arsenide are used in photovoltaic technology. Water-based cells made of conventional crystalline silicon or gallium arsenide are also used and new thin-film technologies are advancing as a result of active research and development initiatives [28, 29].

Concentrated Solar power plants

In addition to using concentrated solar power to generate electricity, technologies for concentrated solar thermal (CST) energy are being used to meet these heating needs. The latter uses large mirrors to concentrate solar energy, which is then used to produce thermal energy to run a steam engine [30].

Four different types of CSP technologies are currently available

- Parabolic troughs, which focus sunlight onto a working liquid in a tube
- Fresnel mirrors use a number of flat reflectors to focus sunlight from the sun onto a receiver tube.

- Power towers, which are field-positioned collections of tens of thousands of sun-tracking reflecting mirrors used to focus solar radiation at a specific location.
- Solar dish collector (which focuses Solar Thermal energy to concentrate power onto an isolated spot located above a reflector dish) [31].

2. Wind Energy

Wind power is one of the most readily available and cleanest energy sources. One of the most readily available is wind energy. It is produced by air movement, whereby the wind spins massive wind turbines to produce power.

Wind energy is sustainable and does not release carbon emissions as a by-product. Wind does not produce a constant, uninterrupted supply of power because wind intensity varies. Because wind speeds vary seasonally, wind farms necessitate a large amount of land where no other activity can take place. Birds could be harmed by wind turbines. Wind turbines that are somewhat higher off the ground than others could produce more energy.

Wind Power Plants

Wind energy is considered to be the most advanced source of clean or renewable energy in terms of both technology and economics [32]. Wind energy systems are ranked second in IRENA's annual reports. With the exception of hydropower, which is a mature technology, solar systems have attracted a lot of attention as a new green energy sources in recent years. This is because it has significant environmental and climate change benefits, as well as numerous other advantages for the environment, business, and society [33]. A wind turbine transforms the rotational energy of the wind into electrical power for a generator. Wind turbines use aerodynamically shaped blades to capture wind energy and turn it into rotational mechanical power. Aerofoils are used in wind turbine blades to produce mechanical power [34]. In poor nations with limited access to electricity, wind energy is the main energy source. Even in remote, challenging, and mountainous locations, it is a practical source of electricity that can be installed and transferred [35]. Wind energy production never runs out of resources and never gets more expensive. Millions of tonnes of pollution and billions of barrels of oil could be avoided because of the electricity generated by these systems [36]. The first wind turbines for generating electricity were built in early 20th century. Technology has evolved progressively since the early 1970s, progress has been made by the decade's end, and wind energy has come back to prominence as one of the most significant sustainable resources for energy [37]. It is difficult for the industry to develop designs that are cost-effective because the kinetic energy of moving air must first be transformed into mechanical energy and then into electrical energy. To complete this conversion, efficient power plants and wind turbines are needed. Theoretically, the quantity of wind kinetic energy that can be extracted increases by a factor of wind speed. However, a turbine only uses a very small amount of the available energy. As a result, minimising energy extraction across the range of wind speeds experienced by wind turbines is given priority in the design of wind turbines (40–50%) [38]. Although wind energy may be useful for minimal power requirements in remote areas, it should be combined with other power production technologies for maximum flexibility to ensure continuity [39].

3. Wave Energy

Water is heavier than air, therefore even a stream that moves slowly can produce a lot of energy. Waves appear on the surface of lakes and oceans as a result of some of the wind's energy being transferred to the water through friction between air and water molecules. Larger waves result from stronger winds. Another sustainable and alternative energy source is wave energy. It is possible to harness the waves up and down motion to produce power. This method of generating energy is still very new and has so far only generated a tiny amount of energy.

4. Biomass

Biomass must have at least 80% of its volume in materials made from living organisms that were collected during the ten years prior to its manufacturing. It is also referred to as "organic matter" and includes anything that is or once was alive. The fact that biomass must be cultivated, gathered, dried, fermented, and burned is a problem.

There are 3 types of biomasses that are as follows:

i) Solid biomass

- Crops like rice, peanuts, maize, and cotton
- Animal waste
- Wood
- Dried and compressed peat
- Straw

Burning wood often costs less than using fossil fuels in a home and delivers heat that is frequently compared to that of a central heating system. Biomass is energy that comes from plant matter and is created by sunlight-driven photosynthesis.

ii) Liquid Biomass

- Vegetable oil is used to make biodiesel, a fuel that can be utilized in existing diesel engines without requiring any hardware modifications. Compared to oil, it is less expensive, sustainable, and non-toxic.
- Ethanol can be produced from sugarcane and corn. It is used as automotive fuel and gasoline additive.
- Biomass can also be used to generate methanol, which is currently made from natural gas.

iii) Gaseous Biomass

- Wood gas
- Biomethane created by the organic degradation of waste or manure from agriculture
- Gasification, which produces carbon monoxide
- Hydrogen produced by the electrolysis of water

Biogas is a gas that may be used to cook food and light dwellings that is created when trash is burned. Many organic materials, including wood, can create gas when subjected to high pressure, high temperature, and anaerobic conditions.

Biomass power plants

Biomass energy is one of the earliest forms of energy. People have used wood to heat their homes and cook their meals for centuries. India has long used a significant quantity of biomass energy. It can be found in firewood, husk, bovine dung cake, and numerous other natural feed sources. Direct application, on the other hand, is less safe because it generates a lot of smoke and ash when biomass is used directly in solid form. The Indian government supports biogas plants as a consequence because they don't produce smoke and are therefore pollution-free.

There are numerous subsidies accessible for building the biogas plant. New gasification technology for biomass has also developed, converting waste into syngas more effectively. Gaseous fuel is produced from biomass [40]. Traditional coal fired power plants can generate electricity from biomass. It is classified as a renewable energy source because its natural supply is not limited. Even though biomass energy shares many characteristics with fossil fuels, it is sustainable and renewable. Biomass can not only be burned to produce energy directly but it can also be used as a feedstock in various processes to produce a range of liquid or gas fuels. Biofuels have the potential to produce heat and power that can be delivered and stored on demand in an energy mix that is highly dependent on erratic energy sources, such as wind. The substantial role that biomass is projected to play in the future is explained by these similarities [41]. To transform biomass feedstock into clean energy fuels, biorefinery and biotransformation technologies must be developed. However, evidence indicates that some environmental effects are caused by biomass power plants. The subsections that follow demonstrate these effects:

- Burning wood and burning waste may result in air pollution
- Biogas
- Liquid biofuels
- Deforestation and farming
- Air emission

Biodiesel and ethanol are examples of biofuels that can be used in place of conventional transportation fuels. Biofuels are often made from sugarcanes and corns. Though, smog and ozone, both of which are hazardous to the environment, these are discharged as a result of evaporation from the ground fuel tanks. To reduce these pollutants, petroleum must be adequately treated before being mixed with ethanol. On the contrary, biodiesel emits less carbon dioxide (CO_2), sulphur oxides, particulate matter, and other harmful unburned hydrocarbons. Compared to petroleum diesel, nitrogen oxide emissions are higher [42]. Jatropha and algae are now also employed in the production of biodiesel [43].

5. Hydroelectricity

Hydropower is defined as the energy generated by moving water that can be transformed into electricity. In hydroelectric dams falling water from rivers or canals is channeled through spin turbines to generate electricity. Hydroelectric dams have major environmental impacts. By sinking enormous tracts of extremely productive forest and wildlife habitat, they harm surrounding habitats. By impeding fish and other species from moving, the dams also harm freshwater ecosystems [44].

Hydropower plants

Hydropower plants may be divided into two categories: those with dams and reservoirs and those without. Hydropower from dams can produce a lot of electricity and has a large amount of storage spaces. Hydropower generated without dams is less expensive. It is advised for simple requirements and is a more environment friendly alternative [45]. In general, hydropower is a clean energy source because it has tremendous potential and creates no waste or greenhouse gas emissions. It can also be increased rapidly and easily. It is a steady energy source with low maintenance costs that foster socioeconomic growth [47], is highly effective, and can adjust to changing environment circumstances [48]. Hydropower can also be used to run equipment, create energy, or do both at once. The usage of mechanical applications is particularly prevalent in limited hydropower facilities, where the electricity produced is utilized to control mechanical equipment for small-scale operations including grinding, pressing, sawing, and milling. The hydroelectric energy production method is claimed to be one of the most affordable in terms of producing power for a well-organized and operated hydropower plant [49], probably because the fuel (falling water) is readily available and cost-free [50]. Water flowing through the hydrological cycle, which is driven by solar energy, generates hydropower. Hydropower can be produced by taking advantage of the gravity-driven movement of water in rivers as it descends from higher altitudes. Based on how they function and the type of water flow, hydropower plants are divided into three categories. The hydrology and topography of the watershed determine the size of storage (reservoir), run-of-river, and pumped storage hydropower plants, which can be small or large. Electricity is produced at a run-of-river hydroelectric power plant using energy. Production is primarily based on the available river flow. Such a hydroelectric facility may have some short-term storage capacity, allowing minor demand profile adjustments. Depending on the local river flow conditions, the profile will alter. As a result, runoff can vary greatly daily, monthly, or seasonally, and generation is dependent on precipitation. Hydropower plants with reservoirs that store water for later use are referred to as storage hydropower facilities. The reservoir lessens the producing station's dependence on inflow fluctuations. Pipelines connect the stations to a downstream tributary that flows into the reservoir. Geography determines the kinds of reservoirs and their designs. Pumped storage hydropower plants are legal; they serve as storage facilities rather than as electricity generators. In such a system, the flow is set aside during the daily peak load time, when power is generated and water is pumped from a lesser reservoir into a higher reservoir, typically during off-peak hours. The plant can deliver huge scale energy storage to consumers. In fact, pumped storage has the greatest capacity. Around the world, electric power storage is already widely available [51].

6. Geothermal energy

Geothermal energy is produced from the earth's internal heat and has the ability to turn a steam turbine into steam, which can then produce electricity. Three miles beneath the surface of the earth, the temperature is 100°C, which is hot enough to boil sufficient water to run a steam powered electric power plant.

Scientists may use geothermal hotspots, which are volcanic phenomena found all over the planet, to obtain this electricity rather than excavating three miles below the earth's surface. Because of the thin mantle at the hotspot, excessive heat from the earth's interior is transferred to the crust.

Electricity may be produced readily at these geothermal hotspots. Here are several examples:

- One method involves injecting hot water into subterranean permeable sedimentary hotspots, where the steam released is used to produce energy. Condensed used steam is released into the permeable sedimentary stream.
- Another device boils water to generate energy while using partially molten volcanic magma at around 650°C.
- The third strategy employs hot dry rock that is simply cemented magma and is still extremely hot.

Water is pumped through the rocks to create steam in order to recuperate this heat.

Renewable Energy Sources how to grow, benefit and invest

Renewable energy has the potential to deliver several societal advantages. Nations have adopted renewable energy policies in addition to reducing carbon dioxide (CO_2) emissions to achieve a number of objectives, such as establishing local environment and health benefits, facilitating access to energy in remote areas, enhancing energy security goals by extending the energy technology portfolio and resources, and fostering social and economic growth through potential employability [52]. 2011 and 2012 saw a rise in the demand for renewable energy. In 2012, the capacity of all renewable energy sources exceeded 1470 GW, an increase of about 8.5% from 2011. A 3% rise in hydropower to an anticipated 990 GW and a 21.5% increase to 480 GW were recorded. Wind power accounted for approximately in 2012, and renewable energy capacity increased by 39%, with solar PV and hydropower each contributing to about 26%. The capacity of wind energy to produce electricity increased 20% globally in 2016. Each year, a larger proportion of new electric capacity is installed globally using renewable energy. Based on various scenarios, projections for the global capacity for renewable energy in 2030 showed that solar PV capacity will increase by 7–25 times and wind power capacity will increase between 4–12 times. CSP is between 20 and 350 times more powerful than biopower, which is between 3 and 5. Based on the real 2011 GW capacity, geothermal is between 4 and 15-fold and hydro is between 30% and 80%.

Non-renewable energy Sources

1. Coal

It takes millions of years for dead plant materials to decompose into peat, which is then heated and compressed under deep burial to become coal [53]. It is a non-renewable source of energy as a result. Fuel is the main application for coal. Heavy metals, sulphur dioxide, nitrogen oxides, and particle pollution, are released by coal-fired power plants, and these pollutants are harmful to human health [54]. Coal burning emits large amounts of sulphate and nitrogen, which can poison rivers and result in haze and fog, as well as greenhouse gas emissions. Mercury produced by coal also accumulates in the fatty tissues of animals and fish, which can be harmful to people who eat these animals and fish.

2. Crude Oil

It develops as a result of massive quantities of decomposing primarily zooplankton, animals, and algae, being suppressed beneath sedimentary rock and being subjected to extreme pressure and heat over the course of millions of years. Oil can be refined to produce goods like jet fuel, diesel, and petrol. The most substantial release of greenhouse gases into the atmosphere occurs during the extraction, refinement and burning of oil. Oil extraction is expensive and frequently harmful to the environment. The surrounding maritime ecology is disturbed by offshore oil exploration and exploitation [55]. Seabirds, animals, shellfish, and other species in coatings can all perish as a result of oil spills.

3. Nuclear Energy

Atomic nuclei serve as a non-renewable source of energy known as nuclear energy. Nuclear fusion or nuclear fission is used to liberate energy. Electricity is produced by nuclear plants using the nuclear fission of the radioactive element Uranium [56]. 17% of the electricity in the world is produced by nuclear power plants. Ionizing radiation is the most commonly used in medical radiography, which uses X-rays to create inside photographs of the human body. Radiation is employed in biology and agriculture to create mutations that result in the emergence of new or improved species, such as in atomic gardening. The handling and eventual disposal of radioactive waste are the principal drawbacks of nuclear power facilities. To prevent material loss and human exposure to radioactive radiation, nuclear reactors require a number of safety precautions.

4. Natural Gas

Is a mixture of hydrocarbon gases that occurs in nature and is primarily composed of methane, though it occasionally also contains trace amounts of other higher alkanes like nitrogen, hydrogen sulphide and helium [57]. It was created from dead plant and animal materials that spent millions of years under the Earth's surface subjected to severe heat and pressure. Energy for heating, cooking and producing of electricity is derived from natural gas. The manufacturing of plastics and other essential organic chemicals for the industry also uses it as a vehicle fuel and a chemical feedstock. It serves as a significant feedstock for the Haber process, which produces ammonia for use in fertilizer manufacture. It is a potent greenhouse gas and releases carbon dioxide into the atmosphere when burnt and contributes to global warming of the earth [58]. Alternative energy is still quite young as of right now. However, due to a mix of political pressure, global ecological disasters, and other factors, this picture is quickly changing and leading to improvements in renewable energy technology.

Novel Synthetic Techniques using a non-conventional source of energy

1. Ultrasound-assisted reactions

Is a green approach used in numerous organic synthesis pathways? Irradiation with high-intensity sound causes acoustic cavitation. Due to its high energy, capacity to disseminate reagents in small particles and ability to speed up reactions, sonochemistry offers a wide range of uses [59]. The true value of ultrasound is seen in its exceptional selectivity and responsiveness augmentation. The most significant contribution to heterogeneous responses is probably made by ultrasound, which selectively accelerates potentially antagonistic pathways. It is well recognized to speed up the reaction process and shorten reaction times. It enhances chemical reactions and mass transfer [60-62]. In addition, it is known to trigger a particular type of reactivity known as "sonochemical switching." When benzoyl bromide was treated with potassium cyanide-impregnated alumina, sonication of the reaction mixture produced benzoyl cyanide whereas heating (without sonication) produced diphenylmethanes. (Scheme 1).

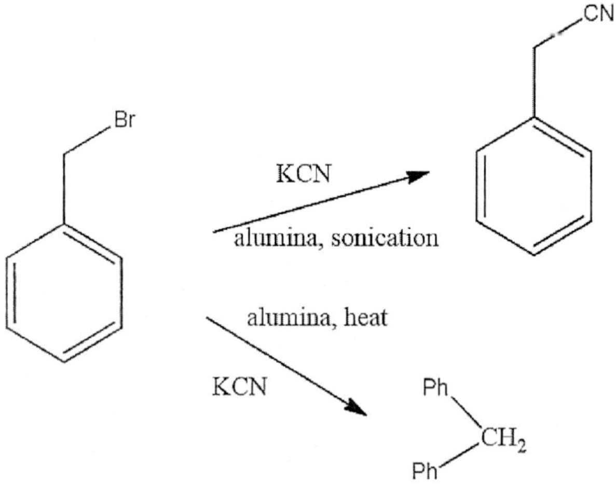

Scheme 1: Sonochemical switching

Ultrasound assisted reactions occurs faster, require cheaper reagents and less extreme physical and chemical conditions as compared to thermal reactions (Scheme 2).

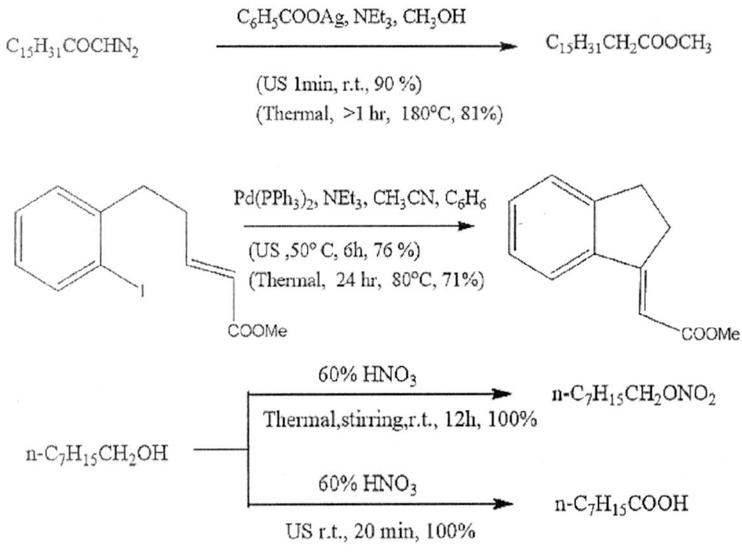

Scheme 2: Thermal versus Ultrasound assisted reactions

2. Microwave-assisted reactions

Microwave-assisted reactions are clean, fast, have high yield, and produce minimum waste [63]. Solvent free microwave aided reactions give researchers the chance to operate in open vessels, which lowers the risk of high pressure and increases the possibility that such reactions may be scaled up [64].

Microwaves are more energy-efficient than traditional heating methods. In the Suzuki reaction, the energy requirement is reduced 85-fold when switching from an oil bath to a microwave reactor [65].

The advantages of microwave irradiation are increased in solvent-free environment because the radiation is directly absorbed by the substrates rather than by the solvents. Increased energy savings have more noticeable implications on yield and selectivity.

One of the instances is the Friedlander coupling condensation of a 2-aminoacetophenone and an acetophenone in the presence of diphenyl phosphate, which produces quinoline derivatives (Scheme 3). Without a solvent, the reactions take place under microwave irradiation in less than 4 minutes. The yield from this microwave-assisted synthesis can reach 85%, but the yield via conventional heating under identical circumstances could not get over 24% [66].

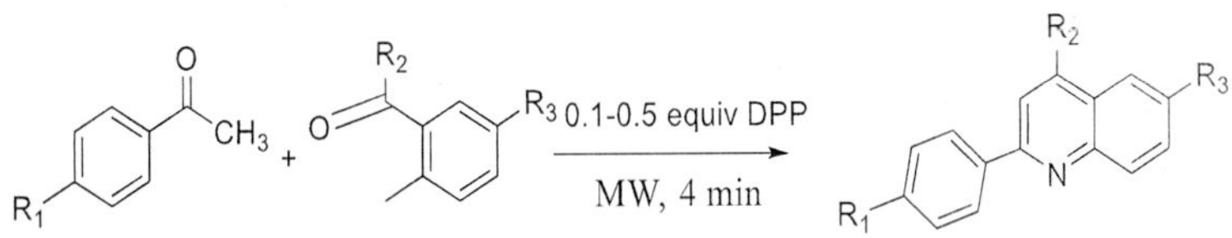

Scheme 3: Microwave assisted preparation of quinoline derivatives in the solvent free conditions

Compared to conventional heating, microwave irradiation has a number of benefits, including homogenous and quick heating (deep interior heating), remarkable reaction accelerations brought on by the heating rate (which are usually impossible to replicate by classical heating), and selective heating [67] shown in scheme 4.

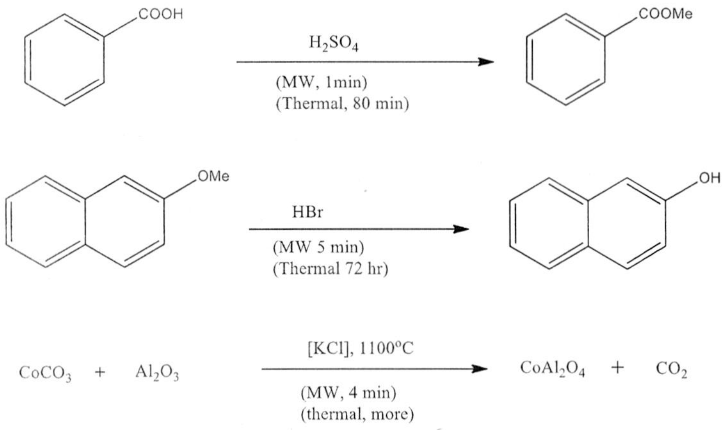

Scheme 4: Thermal versus Microwave assisted reactions

3. Electro Organic Methods

There has been a significant increase in the use of electrochemistry as a synthetic tool in organic chemistry. There are several instances, including the Simon's fluorination, methoxylation, Kolbe reaction, olefin epoxidation, hydrogenation of heterocycles, oxidation of aromatic hydrocarbons, reduction of nitro aromatics and pinacolization.

These encompass a wide range of electrochemical synthesis, including those employed in the pharmaceutical sector, applications. Electrochemical synthesis is very competitive as compare to the conventional methods in organic synthesis due to ease of controlling the solvent and counter-ions, the mild reaction conditions, high selectivity, high yields, simply designed cells, ordinary organic glassware, and use of readily available equipment.

Sacrificial anodes are useful tool for creating metallo-organic compounds through the anodic generation of metal cations and the cathodic generation of organic anions. The organosilicon compounds were successfully synthesized using this method. With very mild conditions and high yields, a stainless-steel cathode and a sacrificial A1 anode can be used to create a wide range of fluorinated organosilicon compounds. Novel electro-organic reaction types that are based on cyclization, coupling and substitution, electrochemically assisted rearrangements and elimination have been discovered [68].

Multicomponent reactions

Multicomponent Reactions (MCRs) are convergent reactions that involve three or more initial components and result in the creation of a new substance in which nearly all or most of the atoms are involved. In MCRs, a product is put together in accordance with a succession of simple chemical processes. A network of reaction equilibriums occurs as a result and they all eventually flow into an irreversible step that produces the product. When one reactant serves as the reaction's media directly, MCRs in water are preferred over those in organic solvents. When the products precipitate out of the reaction mixture, filtration and washing can be used to achieve purification without the need for chromatography or recrystallization [69, 70].

With high diastereoselectivity (diastereomeric excess up to 99%) the synthesis of isoindolinones is 32–67% yields via a multicomponent reaction of Alkyl/aryl-substituted urea, an aromatic aldehyde, and o-phthalaldehyde in the presence of catalytic amounts of trimethyl (Scheme 5) is an example of a MCR [71].

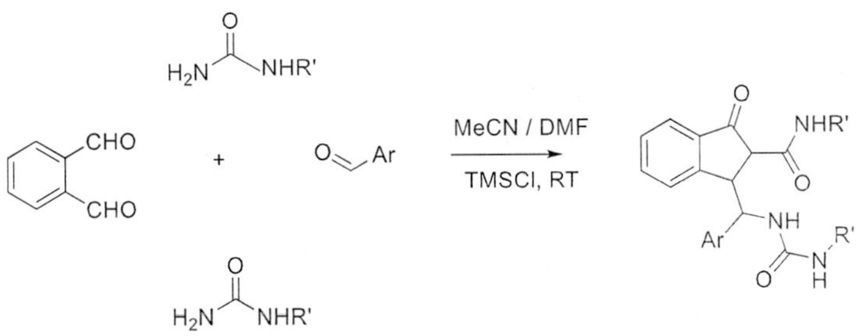

Scheme 5: Synthesis of isoindolinones

Mannich reaction, Strecker reaction and Ugi reaction are all multicomponent reactions (Scheme 6).

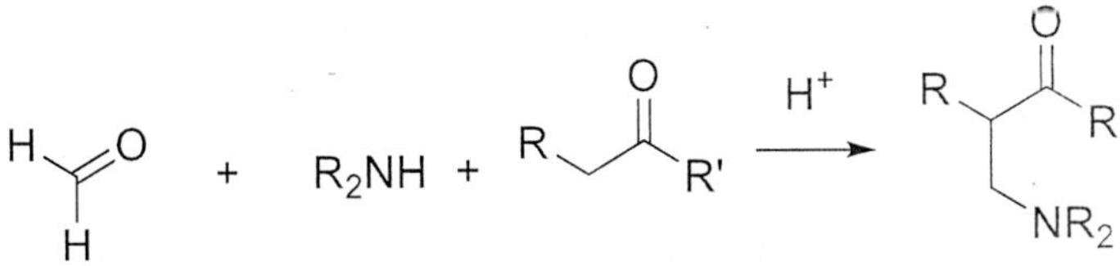

Scheme 6: Mannich reaction

Conclusion

In this study, a thorough assessment of the environment effects of electrical power plants based on RES was done. In this chapter, we look at some potential RES-based power plants, such as wind, solar, hydropower plants, and biomass are all examples considered. Each RES has been found to have an impact on the environment. But with careful choice and regular application, these can be diminished. The adverse environmental consequences are determined by the location, types of fuel/RESs, method of use, size/scale and other factors all play roles as stated in this chapter. For instance, despite the fact that solar photovoltaics have a number of detrimental effects on the environment, it is challenging to reduce their use due to the abundance of solar power plants that are currently in operation. However, by making use of a building's rooftop area, the impact on the land can be diminished. Conventional power generation that depends on fossil fuels is typically viewed as unverifiable over the long term due to a lack of inexhaustible resources and environmental problems caused by emissions. In order to increase the use of renewable energy sources in the electricity sector, numerous efforts are made on a global scale. New options for power generation exist in the form of renewable energy. This chapter provides a current and comprehensive status report as well as a future projection of their benefits, growth, and investment. Furthermore, renewable energy resources should be integrated into smart grids. Keeping all challenges in mind, the system will aid in addressing ever-increasing electric energy demand effectively. The key to environment friendly and economically viable chemical synthesis is therefore the careful application of appropriate synthetic transformations in conjunction with an energy efficient kinetic activation.

References

1. Rahman A, Aziz T, Deeba SR. A time of use tariff scheme for demand side management of residential energy consumers in Bangladesh. Energy Reports. 2021 Nov 1; 7:3189-98.
2. Rodrigues E, Gomes Á, Gaspar AR, Antunes CH. Estimation of renewable energy and built environment-related variables using neural networks–A review. Renewable and Sustainable Energy Reviews. 2018 Oct 1;94:959-88.
3. Ma S, Chen G, Guo M, Zhao L, Han T, Zhu S. Path analysis on CO2 resource utilization based on carbon capture using ammonia method in coal-fired power plants. Renewable and Sustainable Energy Reviews. 2014 Sep 1;37:687-97.
4. Yang Y, Campana PE, Yan J. Potential of unsubsidized distributed solar PV to replace coal-fired power plants, and profits classification in Chinese cities. Renewable and Sustainable Energy Reviews. 2020 Oct 1;131:109967.
5. Chen X, Mauzerall DL. The expanding coal power fleet in Southeast

Asia: Implications for future CO_2 emissions and electricity generation. Earth's Future. 2021 Dec;9(12):e2021EF002257.

6. Gürsan C, de Gooyert V. The systemic impact of a transition fuel: Does natural gas help or hinder the energy transition? Renewable and Sustainable Energy Reviews. 2021 Mar 1;138:110552.
7. Restrepo-Herrera D, Martinez W, Trejos-Grisales LA, Restrepo-Cuestas BJ. A Holistic Approach for Design and Assessment of Building-Integrated Photovoltaics Systems. Applied Sciences. 2023 Jan;13(2):746.
8. Todeschini G. Review of tidal lagoon technology and opportunities for integration within the UK energy system. Inventions. 2017 Jul 28;2(3):14.
9. Shao M, Han Z, Sun J, Xiao C, Zhang S, Zhao Y. A review of multi-criteria decision making applications for renewable energy site selection. Renewable Energy. 2020 Sep 1;157:377-403.
10. Ganiyu SO, Martinez-Huitle CA, Rodrigo MA. Renewable energies driven electrochemical wastewater/soil decontamination technologies: A critical review of fundamental concepts and applications. Applied Catalysis B: Environmental. 2020 Aug 5;270:118857.
11. Graebig M, Bringezu S, Fenner R. Comparative analysis of environmental impacts of maize–biogas and photovoltaics on a land use basis. Solar Energy. 2010 Jul 1;84(7):1255-63.
12. Arvesen A, Hertwich EG. Assessing the life cycle environmental impacts of wind power: A review of present knowledge and research needs. Renewable and sustainable energy reviews. 2012 Oct 1;16(8):5994-6006.
13. Kuriqi A, Pinheiro AN, Sordo-Ward A, Garrote L. Influence of hydrologically based environmental flow methods on flow alteration and energy production in a run-of-river hydropower plant. Journal of Cleaner Production. 2019 Sep 20;232:1028-42.
14. Zhang L, Pang M, Wang C, Ulgiati S. Environmental sustainability of small hydropower schemes in Tibet: An emergy-based comparative analysis. Journal of cleaner production. 2016 Nov 1;135:97-104.
15. Kabir E, Kumar P, Kumar S, Adelodun AA, Kim KH. Solar energy: Potential and future prospects. Renewable and Sustainable Energy Reviews. 2018 Feb 1;82:894-900.
16. Child M, Koskinen O, Linnanen L, Breyer C. Sustainability guardrails

for energy scenarios of the global energy transition. Renewable and Sustainable Energy Reviews. 2018 Aug 1;91:321-34.

17. Kabir E, Kumar P, Kumar S, Adelodun AA, Kim KH. Solar energy: Potential and future prospects. Renewable and Sustainable Energy Reviews. 2018 Feb 1;82:894-900.

18. CD O. Linking renewable energy to rural development. OE CD Green Growth Studies.

19. Bergmann A, Hanley N, Wright R. Valuing the attributes of renewable energy investments. Energy policy. 2006 Jun 1;34(9):1004-14.

20. Del Río P, Burguillo M. Assessing the impact of renewable energy deployment on local sustainability: Towards a theoretical framework. Renewable and sustainable energy reviews. 2008 Jun 1;12(5):1325-44.

21. Waldenström C, Ferguson R, Sundberg C, Tidåker P, Westholm E, Åkerskog A. Bioenergy from agriculture: Challenges for the rural development program in Sweden. Society & Natural Resources. 2016 Dec 1;29(12):1467-82.

22. Dai H, Xie X, Xie Y, Liu J, Masui T. Green growth: The economic impacts of large-scale renewable energy development in China. Applied energy. 2016 Jan 15;162:435-49.

23. Farret FA, Simões MG. Integration of alternative sources of energy. Piscataway, NJ, USA: IEEE press; 2006 Jan 17.

24. Devabhaktuni V, Alam M, Depuru SS, Green II RC, Nims D, Near C. Solar energy: Trends and enabling technologies. Renewable and Sustainable Energy Reviews. 2013 Mar 1;19:555-64.

25. Liu Y, Zhang M. Unpacking stereotypes about sustainable energy: Knowledge, policy, and public misperceptions of solar energy and coal in China. Energy Research & Social Science. 2021 Jan 1;71:101832.

26. Hosenuzzaman M, Rahim NA, Selvaraj J, Hasanuzzaman M, Malek AA, Nahar A. Global prospects, progress, policies, and environmental impact of solar photovoltaic power generation. Renewable and sustainable energy reviews. 2015 Jan 1;41:284-97.

27. Mohanty P, Muneer T, Gago EJ, Kotak Y. Solar radiation fundamentals and PV system components. InSolar Photovoltaic System Applications: A Guidebook for Off-Grid Electrification 2015 Sep 17 (pp. 7-47). Cham: Springer International Publishing.

28. Jeon NJ, Lee J, Noh JH, Nazeeruddin MK, Gratzel M, Seok SI. Efficient

inorganic–organic hybrid perovskite solar cells based on pyrene arylamine derivatives as hole-transporting materials. Journal of the American Chemical Society. 2013 Dec 26;135(51):19087-90.

29. Alharbi FH, Kais S. Theoretical limits of photovoltaics efficiency and possible improvements by intuitive approaches learned from photosynthesis and quantum coherence. Renewable and Sustainable Energy Reviews. 2015 Mar 1;43:1073-89.

30. Philibert C. Solar energy perspectives. OECD; 2011.

31. Nazir MS, Ali N, Bilal M, Iqbal HM. Potential environmental impacts of wind energy development: A global perspective. Current Opinion in Environmental Science & Health. 2020 Feb 1;13:85-90.

32. Dai K, Bergot A, Liang C, Xiang WN, Huang Z. Environmental issues associated with wind energy–A review. Renewable energy. 2015 Mar 1;75:911-21

33. Blaabjerg F, Teodorescu R, Chen Z, Liserre M. Power converters and control of renewable energy systems. InProc. 6th Int. Conf. Power Electron 2004 Oct 18 (Vol. 1, pp. 1-20).

34. Singh S, Bhatti TS, Kothari DP. Indian scenario of wind energy: problems and solutions. Energy Sources. 2004 Jul 1;26(9):811-9.

35. Bellarmine GT, Urquhart J. Wind energy for the 1990s and beyond. Energy Conversion and Management. 1996 Dec 1;37(12):1741-52.

36. Kaygusuz KA. Wind power for a clean and sustainable energy future. Energy Sources, Part B. 2009 Jan 6;4(1):122-33.

37. Pryor SC, Barthelmie RJ. Climate change impacts on wind energy: A review. Renewable and sustainable energy reviews. 2010 Jan 1;14(1):430-7.

38. Lobão JA, Devezas T, Catalão JP. Reduction of Losses in the Conductors of a Sustainable Street Lighting as a Technique for Managing Energy Consumption and CO2 Emissions.

39. Nagrath IJ, editor. Power system engineering. Tata McGraw Hill India; 2007.

40. Bhattacharyya SC. Energy access problem of the poor in India: Is rural electrification a remedy?. Energy policy. 2006 Dec 1;34(18):3387-97.

41. Hall DO, Scrase JI. Will biomass be the environmentally friendly fuel of the future?. Biomass and Bioenergy. 1998 Oct 11;15(4-5):357-67.

42. Fiorentino G, Ripa M, Ulgiati S. Chemicals from biomass: technological

versus environmental feasibility. A review. Biofuels, Bioproducts and Biorefining. 2017 Jan;11(1):195-214.

43. Williams TO, Fernández-Rivera S, Kelley TG. The influence of socioeconomic factors on the availability and utilization of crop residues as animal feeds.

44. Ellabban O, Abu-Rub H, Blaabjerg F. Renewable energy resources: Current status, future prospects and their enabling technology. Renewable and sustainable energy reviews. 2014 Nov 1;39:748-64.

45. Halkos GE, Gkampoura EC. Reviewing usage, potentials, and limitations of renewable energy sources. Energies. 2020 Jun 5;13(11):2906.

46. Owusu PA, Asumadu-Sarkodie S. A review of renewable energy sources, sustainability issues and climate change mitigation. Cogent Engineering. 2016 Dec 31;3(1):1167990.

47. Okot DK. Review of small hydropower technology. Renewable and Sustainable Energy Reviews. 2013 Oct 1;26:515-20.

48. Mohtasham J. Renewable energies. Energy Procedia. 2015 Aug 1;74:1289-97.

49. Kaunda CS, Kimambo CZ, Nielsen TK. Hydropower in the context of sustainable energy supply: a review of technologies and challenges. International Scholarly Research Notices. 2012;2012.

50. Brown A, Müller S, Dobrotkova Z. Renewable energy: Markets and prospects by technology. IEA information paper. 2011 Nov.

51. Bildirici ME. Economic growth and biomass energy. Biomass and bioenergy. 2013 Mar 1;50:19-24.

52. Angelou N, Elizondo Azuela G, Banerjee SG, Bhatia M, Bushueva I, Inon JG, Jaques Goldenberg I, Portale E, Sarkar A. Global tracking framework. The World Bank; 2013 May 28.

53. Blander M, Sinha S, Pelton A, Eriksson G. Calculations of the influence of additives on coal combustion deposits. Argonne National Lab., IL (USA); 1989 Jan 1.

54. Hendryx M, Zullig KJ, Luo J. Impacts of coal use on health. Annual review of public health. 2020 Apr 1;41:397-415.

55. Rana S. Facts and Data on Environmental Risks—Oil and Gas Drilling Operations. InSPE Asia Pacific Oil and Gas Conference and Exhibition 2008 Oct 20. OnePetro.

56. Abadi VM. Nuclear energy, the past and the future challenges in competition with renewable energy in Iran and the world. American Journal of Renewable and Sustainable Energy. 2016;2(2):6-11.

57. Khamis AS. Design Plant Regenerative Process of Acid Gas Removal.

58. Kobayashi M, Nagasawa T, Yamada H. Enzymatic synthesis of acrylamide: a success story not yet over. Trends in biotechnology. 1992 Jan 1;10:402-8.

59. Kobayashi M, Shimizu S. Nitrile hydrolases. Current opinion in chemical biology. 2000 Feb 1;4(1):95-102.

60. Cintas P, Luche JL. Green chemistry. The sonochemical approach. Green Chemistry. 1999;1(3):115-25.

61. Mason TJ. Ultrasound in synthetic organic chemistry. Chemical Society Reviews. 1997 Jan 1;26(6):443-51.

62. Cravotto G, Cintas P. Power ultrasound in organic synthesis: moving cavitational chemistry from academia to innovative and large-scale applications. Chemical Society Reviews. 2006;35(2):180-96.

63. Luche JL, Cintas P, Luche JL. Organometallic sonochemistry. Synthetic Organic Sonochemistry. 1998:167-234.

64. Tuulmets A, Kaubi K, Heinoja K. Influence of sonication on Grignard reagent formation. Ultrasonics sonochemistry. 1995 Jan 1;2(2):S75-8.Einhorn, C., J. Einhorn, M. J. Dickens, and J. L. Luche. "Organic sonochemistry—some illustrative examples of a new fundamental approach." *Tetrahedron letters* 31, no. 29 (1990): 4129-4130.

65. Lidström P, Tierney J, Watheyb B, Westmana J. Microwave assisted organic synthesisÐa review. Tetrahedron. 2001;57:9225-83.

66. Kidwai M. Dry media reactions. Pure and Applied Chemistry. 2001 Jan 1;73(1):147-51.

67. Baghurst DR, Mingos DM. A new reaction vessel for accelerated syntheses using microwave dielectric super-heating effects. Journal of the Chemical Society, Dalton Transactions. 1992(7):1151-5.

68. Doble M, Rollins K, Kumar A. Green chemistry and engineering. Academic Press; 2010 Jul 27.

69. Moond M, Singh S, Sangwan S, Rani S, Beniwal A, Rani J, Kumari A, Rani I, Devi P. Phytofabrication of Silver Nanoparticles Using Trigonella foenum-graceum L. Leaf and Evaluation of Its Antimicrobial

and Antioxidant Activities. International Journal of Molecular Sciences. 2023 Feb 9;24(4):3480.

70. Zhu J, Bienaymé H, editors. Multicomponent reactions. John Wiley & Sons; 2006 Mar 6.

71. Wan JP, Zhou J, Mao H, Pan YJ, Wu AX. A new diastereoselective multicomponent, one-pot strategy for the synthesis of 3-substituted isoindolinones via efficient C–C bond formation. Tetrahedron. 2008 Dec 1;64(49):11115-23.

Chapter - 3

Catalysts and Reagents

Abstract

The creation of chemical products and methods that lessen or do away with the usage and manufacturing of hazardous compounds is known as "green chemistry." Green chemistry has obtained popularity as a fresh method of environmentally sound protection over the past ten years. Catalysis has established itself as a crucial instrument in the fight against pollution. While catalysis has long been used to boost productivity, yield, and selectivity, it is now acknowledged that it also helps to achieve a variety of green chemical objectives. In order to lessen or totally eliminate the use of dangerous compounds, green chemistry applies a set of principles to the development, manufacture, and use of chemical products. Academic research on the topic of green solvents does not concentrate on the industries that use solvents the most or the solvent types that the scientific community believes have the best chance of reducing solvent-related environmental damage. Without industry, modern civilization would not be able to function, and economic development is naturally accompanied by rubbish generation. Waste is something that is thrown away without expecting to be compensated for its fundamental value. When these wastes are poorly handled, warehoused, transported, disposed of, or managed, they could constitute a threat to the environment or human health (air, water, or soil,). Given this, it is crucial to handle hazardous wastes, including their disposal, in a practical and eco-friendly way, and suggestions are provided for improving current procedures. Solid trash is the sort of garbage that has the greatest environmental impact out of all the others. The current paper covers the following topics: waste characterization, the nature of wastes, health and environmental effects of waste management practices, industries that produce waste, steps towards design, planning, and development of models for effective management of hazardous waste, regulations, approaches, and treatment for disposal of hazardous waste.

Keywords: Green chemistry, catalysis, green solvent

Introduction

Like organic chemistry, green chemistry presents a wealth of opportunities for insightful analysis, discussion, and critique. The goal of developing new synthesis techniques is to enhance or replace existing non- or less-green technologies [1].

We must create new, more environment friendly chemical goods and processes in order to attain the "triple bottom line". Catalysis played a notable role in the industry's success in the 20th century, and it will play a similar role in the new, more environment friendly industry of the 21st. [2]. By replacing reagents or enabling more efficient processes, for example, catalysis can not only contribute to the greening of chemical processes but also serve as a catalyst for this change by lowering costs and environmental effects [3].

Catalytic reagents that are as selective as possible outperform stoichiometric reagents in terms of efficiency. The catalysis principle promotes the use of biodegradable catalysts, which use less energy, don't include organochlorine compounds, and conserve water while producing less wastewater, in order to safeguard the environment [4].

Large volumes of flammable and volatile organic solvents are employed in several industrial processes' reaction systems and separation phases, which significantly affects a process's economic and environmental performance [5]. Designing novel, environmentally acceptable, and tunable solvents that would be used to satisfy both technological and economic objectives is thus a major topic of research in the development of green technology [6]. Benzene, chlorinated organic solvents, and other frequently used solvents are among the numerous documented carcinogens and environmental hazards.

Hence, it is important to limit the use of solvents, and when it is unavoidable, the solvent should be non-flammable, environmentally friendly, non-toxic, and, like water. To lessen the harm these compound cause to the environment, organic solvents must be replaced. One effect is the requirement to avoid using organic solvents whenever possible. Optimization, minimization, and recycling of solvents should be attempted when elimination is not practical. The best solvents are those that are stable, affordable, easily accessible, and have minimal negative influence on the environment. Ionic liquids at ambient temperature, subcritical fluids, solvents derived, and supercritical from renewable and natural sources be distinctive as the most viable techniques for contemporary solvent innovation among the proposed solvents [7]. This article gives a concise summary of current information about these solvents with a focus on their characteristics, uses, and potential as industrial solvents that are truly green [8].

Due to the large-scale production of various chemicals and energy as well as other development activities like health care, urbanization, and agriculture during the past 40 years, enormous amounts of waste have been permitted to enter the environment. These wastes have taken the form of solids, liquids, and gases. All living things, including humans, are gravely threatened by the substantial number of potential environmental risks included in these wastes. It is crucial to effectively and practically manage such hazardous trash if we are to protect our ecology. On-site regulations are followed in compliance with regulatory requirements for the disposal of wastes that must be unavoidably discharged into the atmosphere as emanations. With hazardous solid wastes, the issue is more complicated, and long-standing storage, disposal, or use inside the buildings is not practical. The current article primarily addresses issues with industrialized solid waste, as well as silt and slurry from ETPs and other polluted areas, as these wastes require special collection and treatment [9]. The heterogeneity of the management problem makes it more difficult. Hazardous waste treatment is an alternative to landfilling that involves recycling or eliminating the trash. To find efficient and cost-effective waste management methods, there is a development process that is rife with experimentation. When treating hazardous waste, physical techniques including filtration, mechanical collection, forming, sedimentation, centrifugation, leaching, coagulation, etc. are first used to reduce the volume of the material. Then, other chemical treatment techniques including neutralization, oxidation, chemical precipitation, etc. are used [10].

Catalysts

In general, catalysis plays a substantial role in increasing the effectiveness and financial viability of manufacturing processes. Catalysts mostly have three key features, i) The energy input necessary for a process is decreased as a result of catalytic reagents that lower the energy of the transition state, ii) Many catalysts are needed, when it comes to biocatalysts, there are significantly fewer catalysts (often enzymes) required than there are reactants, iii) Catalyst regeneration and reversibility are beneficial for environment friendly processes [11].

Atom efficiency, commonly referred to as atom economy, is a key idea in green chemistry and can be addressed using catalysis. The increasing environmental consciousness has caused a paradigm shift in how we assess the effectiveness of processes. Traditional chemical yields are no longer the main focus; instead, the atom-economy (or efficiency) (AE) factor and environmental quotient (EQ) parameters are now being considered. By viewing waste as everything created beside the desired product, the simplest definition of the AE factor is the quantity of waste produced per kilogram of product. Calculating two potential roots for a given product quickly using the AE method. To calculate it, divide the molecular weight of the intended product by the total molecular weight of all the chemicals created during the processes described in the stoichiometric equation. Some writers refer to it as the total number of reactant atoms that are changed into atoms of the intended product over the course of a reaction. Compared to the catalytic method, processes that are less atom-efficient use stoichiometric reagents [12].

For instance, classic oxidations utilizing oxidants like permanganate or chromium reagents enhance atom economies when swapped out for cleaner-catalyzed oxidation (Scheme 1).

$$3\,PhCH(OH)CH_3 + 2CrO_3 + 3H_2SO_4 \longrightarrow 3PhCOCH_3 + Cr_2(SO4)_3 + 6H_2O$$

Scheme 1: Jones oxidation of alcohol

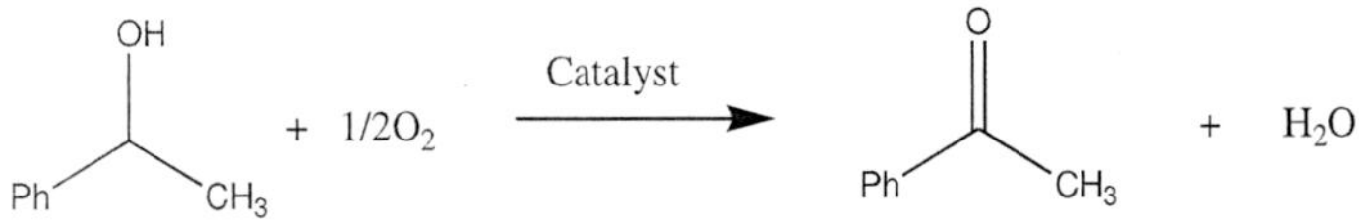

Scheme 2: Oxidation of alcohol

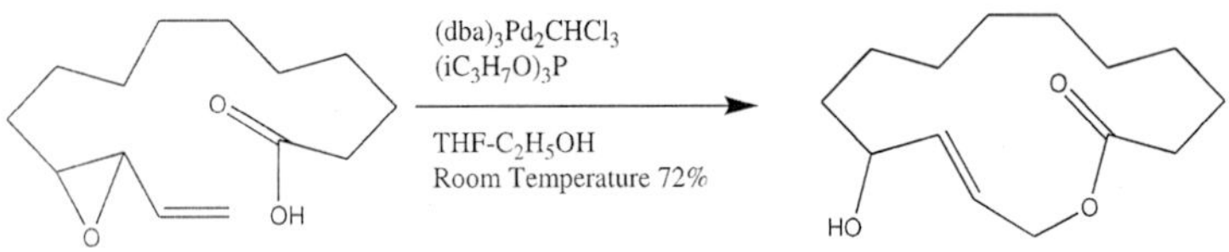

Scheme 3: Macro lactone Synthesis

The atom efficiency for the oxidation of a secondary alcohol to a ketone shown in scheme 1 using stoichiometric reagents is 42%. The atom efficiency is increased by more than twice, to 87%, by the same product produced by catalytic oxidation [13] (Scheme 2).

Trost and colleagues (Trost, 1998) carried out an allylic alkylation process using a range of palladium catalysts. It is entirely atom-efficient (100%) to produce macro lactones from corresponding carboxylic acids with the help of a palladium complex (Scheme 3). The reaction is another illustration of how catalysis saves energy because it takes place at room temperature [14].

In the past, six stoichiometric steps were required to produce ibuprofen, and the atom efficiency was only about 40%. Advil TM and Matrix WM were the brand names used for its marketing. Three phases make up a new catalytic procedure with an atom efficiency of 80% that BHC (BHC, 1997) developed for the synthesis of ibuprofen. Even though the process's use of the hazardous chemical HF is a downside, HF is recovered with a 99.9% efficiency rate. This is a great illustration of a green method that has been made into a product (Scheme 4).

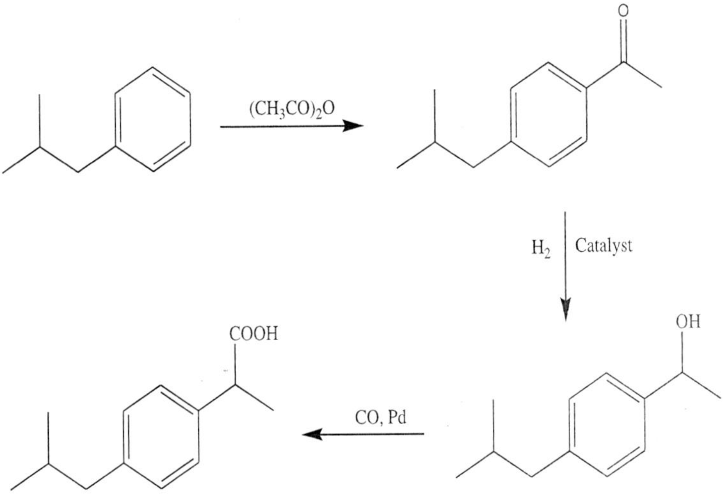

Scheme 4: Green synthesis of Ibuprofen

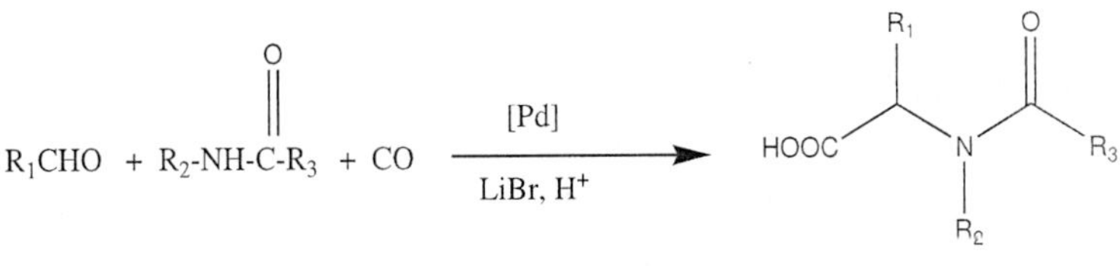

$[Pd] = PdBr_2$ (or) $(Ph_3P)_2PdBr2$

Scheme 5: Amido carboxylation

A classic illustration of a catalysed carboxylation reaction is the procedure depicted in Scheme 4 and 5, sometimes referred to as the Hoechst-Celanere process [15]. The conversion of a CO, an amide, and an aldehyde to an acylamino acid occurs elegantly in one step during carboxylation processes, which are typically 100% atom-efficient (Scheme 5).

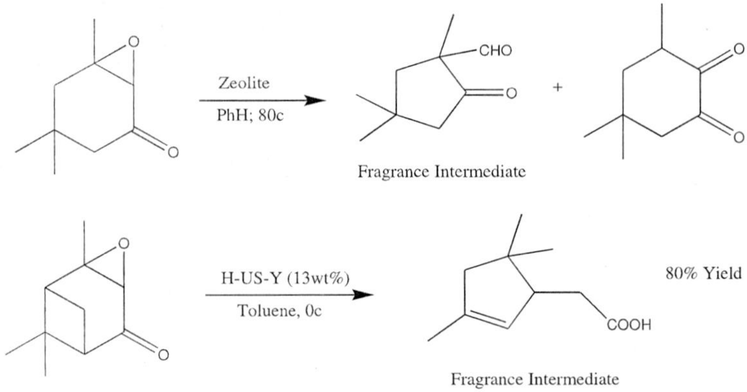

Scheme 6: Reaction in which Zeolites and clay as a catalyst

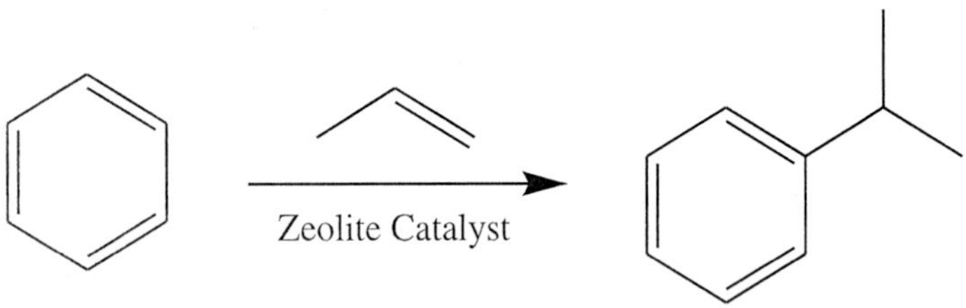

Scheme 7: Synthesis of cumene

It is crucial that zeolites are used in the production of cumene. Cumene has generated annually in excess of 7 million metric tonnes worldwide (Scheme 6). The method that was previously employed involves alkylating benzene over a solid catalyst made of aluminium chloride or phosphoric acid. Manufacture of cumene, U.S. Patent 4008290. The nature of both catalysts is poisonous. Using the less hazardous carozine zeolite catalyst, the Mobil/Badger cumene method (Mobil Technology Co., 1997) was developed. It concurrently satisfies many requirements of green chemistry and produces a lesser amount of waste and uses a smaller amount energy than the older catalysts [16] (Scheme 7).

The release of hazardous compounds into the environment is a significant sustainability concern. For instance, solvents are used often in government, commercial, and university laboratories. Industrial solvents are employed in the chemical process at several stages of operation, including crystallization, reaction (reaction medium), separation, degreasing, reactants, washing, and carrier. Several of the frequently used solvents are hazardous air pollutants (HAPs), poisonous, flammable, and/or volatile organic compounds (VOCs). They also provide significant risks to process safety, human and environmental toxicity, and waste management, among other environmental, health, and safety (EHS) problems. In the Novartis method, a chiral metal compound acts as a catalyst to create the optically active herbicide(s)-metachlor (Blaser and Spindler, 1998) (Scheme 8). An important step in the procedure is the asymmetric hydrogenation of a prochiral imine, which is catalysed by an iridium(I) complex of the asymmetric ferrocenyl diphosphine. This step's substrate/catalyst ratio is 750,000 and it has a high turnover rate, making the method commercially viable [17].

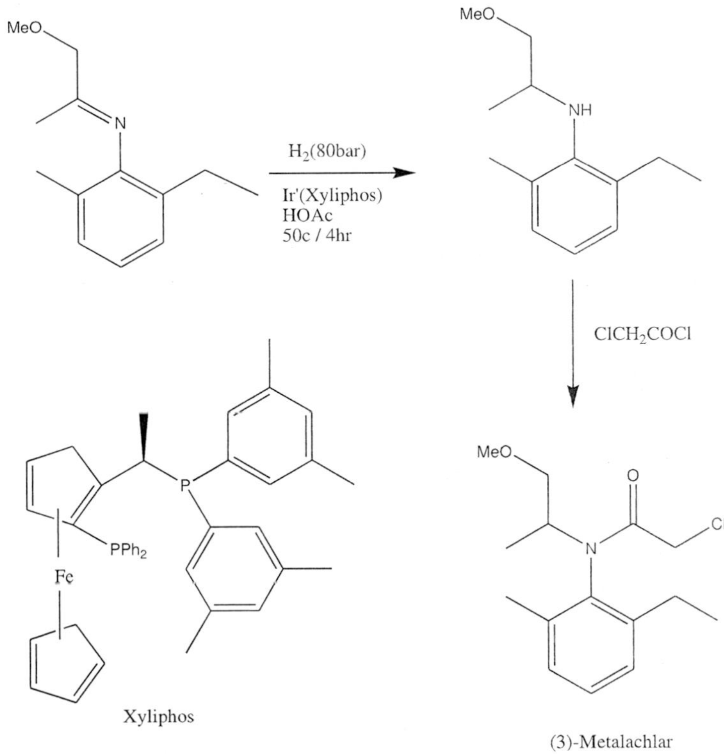

Scheme 8: Process in which synthesis of the chiral herbicide

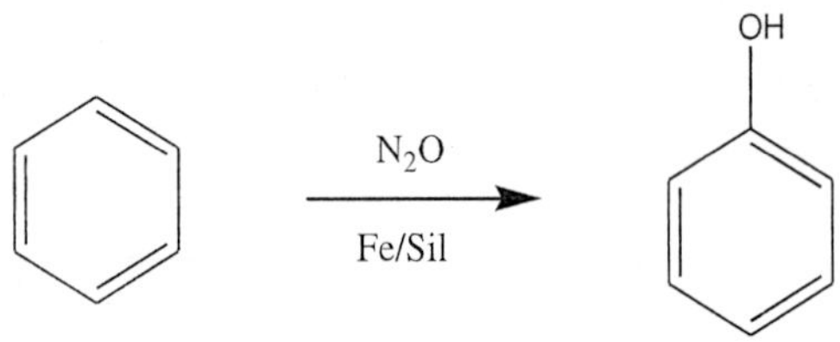

Scheme 9: Synthesis of phenol from benzene

In partnership with the Boreskov Institute of Catalysis in Russia, Solutia (USA) developed a one-step method to convert benzene into phenol utilising nitrous oxide as the oxidant (Scheme 9). A by-product of Solutia's adipic acid production is nitrous oxide, a greenhouse gas. The most favoured catalysts are acidified iron-containing ZSM-5 and ZSM-11 zeolites or 100:1 silica/alumina materials that contain 0.45 weight percent iron (III) oxide. The catalyst has a 3–4 days half-life and can be replenished by blowing hot air through the bed.

Enichem has created a method for producing cumene using zeolite and benzene. The waste produced by poly alkylates using the conventional chemical method is decreased by this catalytic process (Scheme 10).

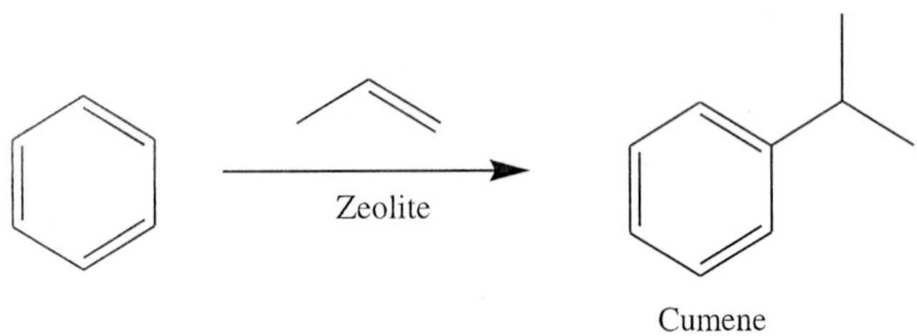

Scheme 10: Synthesis of cumene from benzene.

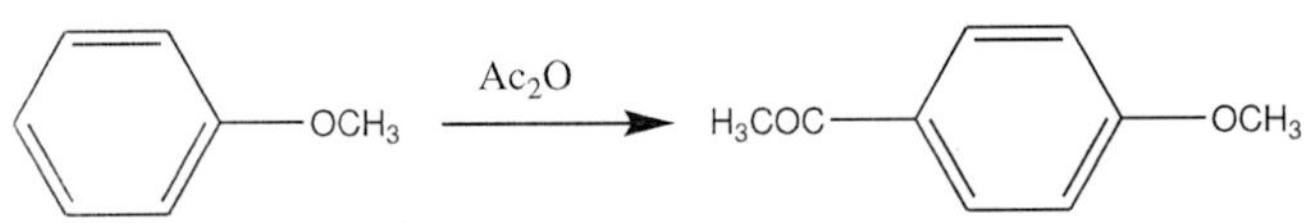

Scheme 11: p-methoxy acetophenone produce from methoxy-benzene.

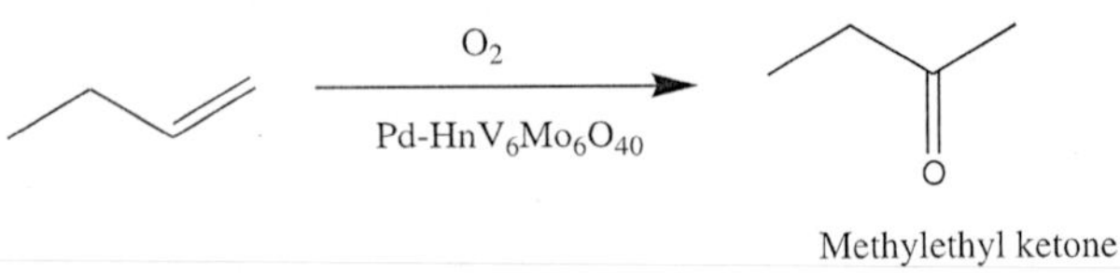

Scheme 12: From ethylene and butylenes produce methyl ethyl ketone (MEK).

By employing clay as the catalyst in the Rhodia process, it is possible to produce p-hydroxy acetophenone from methoxybenzene without requiring hazardous substances like BF_3 and $AlCl_3$ (Scheme 11). The catalytic process for producing methyl ethyl ketone (MEK) from ethylene and butylene uses the oxides of palladium, molybdenum, and vanadium, as catalysts (Scheme 12). Due to the usage of chlorinated chemicals in the initial process, there was a significant volume of chlorinated waste that presented a number of challenges during disposal [18].

Green Solvents

The environment friendly solvents known as "green solvents" or "bio solvents" are produced when agricultural products are processed. As eco-friendlier substitute for petrochemical solvents, green solvents were created (Figure 1).

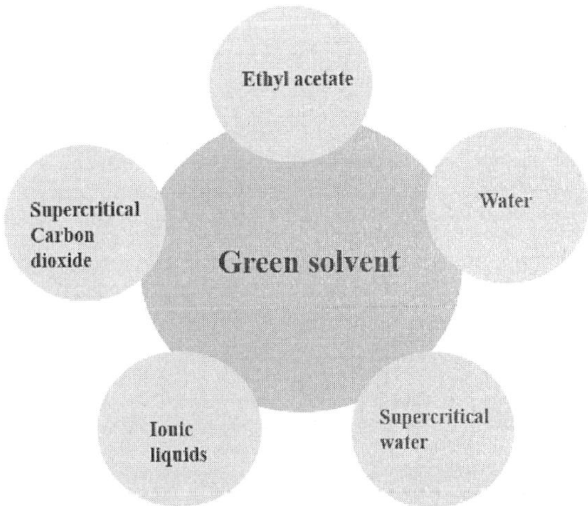

Fig 1: Green solvents

1. Ethyl lactate

It is green solvent which derived from the process of corn. It is a widely used solvent in paint and varnish production and has many attractive benefits, including non-ozone depleting, non-carcinogenic, easy to process, 100% biodegradable, and non-corrosive. The low surface tension, strong resolution, low vapor pressure, and high boiling point of this solvent make it a good solvent for the coatings industry. Ethyl lactate replaces solvents such as acetone, toluene, and xylene, making the workplace safer. Ethyl lactate can also be used as a tremendous cleaner in polyurethane production.

Because ethyl lactate has excellent cleaning performance, it can be used to thoroughly scrub diversification of metal surfaces and remove solid fuels, grease, and adhesives. Ethyl lactate is utilized in palladium-catalyzed Suzuki-Miyaura reactions as a co-solvent of water [19]. Results from ethyl lactate-mediated reactions were shown to be superior to those obtained with more traditional polar organic solvents including dioxane, DMF, and DMSO. (Scheme 13).

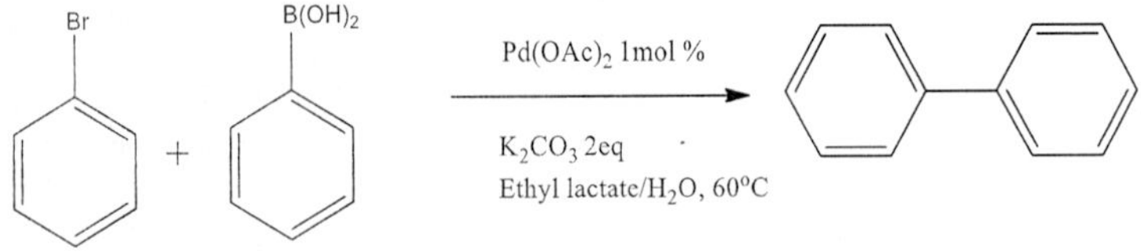

Scheme 13: Manufacture of biphenyl from Suzuki-Miyaura in ethyl lactate/water.

2. Water

Water is universal and the greenest solvent is in a pure state. The use of water as the solvent for organic reactions can avoid the issues with solvent toxicity and disposal. The use of water as a solvent in synthetic chemistry enables the production of compounds at a lower cost and with lower risk. Compared with organic solvents, the Diels-Alder reaction in water proceeds at a higher rate and has higher endo selectivity and exoselectivity. In the Diels-Alder reaction, water as a solvent can promote an endo-transition state. The acceleration of water is attributed to various factors, including hydrophobicity in addition to hydrogen bonds between water molecules and reagents [20]. (Scheme 14).

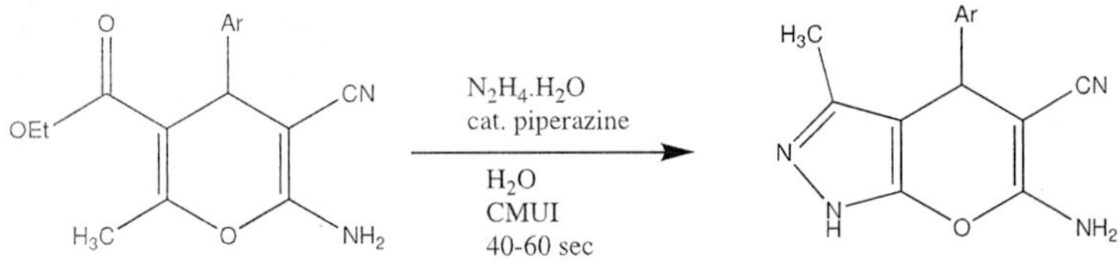

Scheme 14: Diels-Alder reaction.

i) In water the proportion of O vs C allylations of phenoxide ions was demonstrated by Breslow and colleagues, can be determined by hydrophobic interactions [21]. Hydrophobic borohydrides have been used to manage the regioselective reduction of naturally occurring sulphated steroids. When the reaction was conducted in water with $LiBH_4$, there was a large amount of selectivity (87:13) for the naturally more reactive 17-keto group. Due to efficient hydrophobic packing at the 6-keto group, an unexpected turn around of selectivity was seen during the reduction with $LiC_6F_5BH_3$ in 4M $LiCl/D_2O$ [22].

ii) Hydrophobic oxidizing agents can be used to selectively promote the epoxidation of olefins. When hydrophobic oxaziridinium salt is used, the selectivity of epoxidation of cinnamic acid and crotonic acid derivatives is drastically increased [23].

3. Ionic liquids

Organic salts that are liquid at or below 100 °C are referred to as ionic liquids, also referred to as RTILs (room temperature ionic liquids). They typically include an organic cation and a polyatomic inorganic anion. Ionic liquids are mainly notable for having extraordinarily low vapor pressure [24]. Molten salts, neoteric solvents, ionic fluids, and designer solvents are all terminology used to describe ionic liquids. Ionic liquids are composed of a larger organic cation with nitrogen and a smaller inorganic anion. A salt with a low melting point is produced as a result of the asymmetry's reduction of the crystalline structure's lattice energy. The multi-component ionic liquids can be formed by mixing simple liquid salts (monoanionic and cation) with different inorganic salts. By selecting anions and cations, it is conceivable to create an ionic liquid with the favored features for a particular application. Ionic liquids can be altered to produce properties that are particular to some chemical processes, such as miscibility, density, a particular melting temperature, hydrophobicity, viscosity, etc (Figure 2).

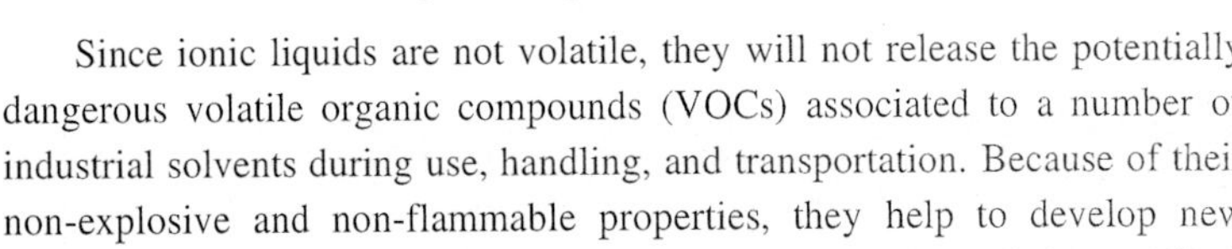

Fig 2: Examples of ionic liquids

Since ionic liquids are not volatile, they will not release the potentially dangerous volatile organic compounds (VOCs) associated to a number of industrial solvents during use, handling, and transportation. Because of their non-explosive and non-flammable properties, they help to develop new reactions, safer and more environmentally friendly processes (Scheme 15).

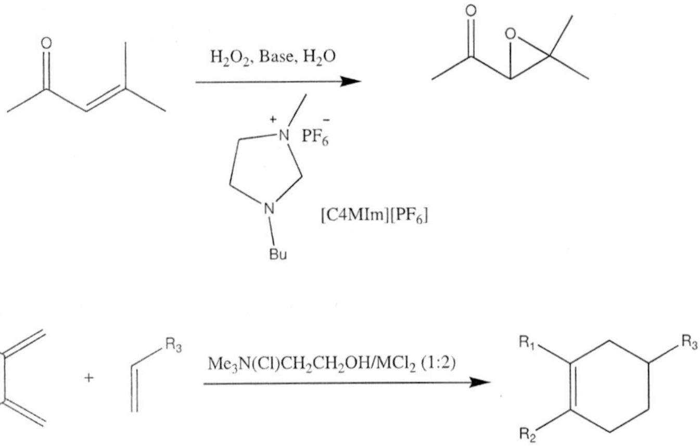

Scheme 15: Diels-alder reaction in an ionic liquid.

They are suitable for various chemical reactions and processes, such as Diels alder, hydrogenation, biocatalysis reactions (e.g. transesterification and hydrolysis), Stereoselective halogenations and many more [25] (Scheme 16).

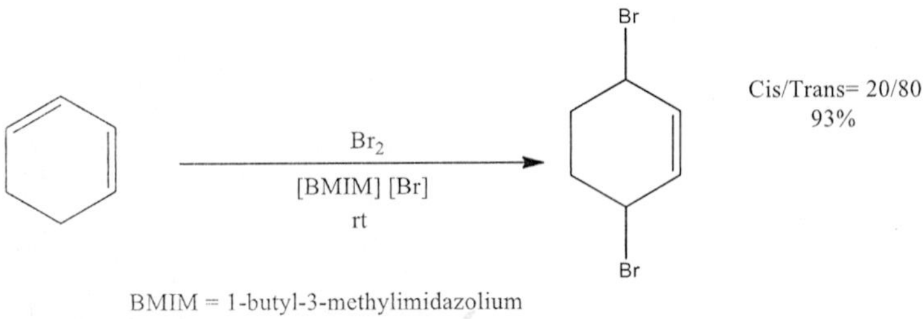

Scheme 16: Representative reaction for stereo selective halogenation

Ionic liquids are used to separate biologically generated raw materials from fermentation-based liquids like ethanol, acetone, or butanol. Ionic liquids performed better than $AlCl_3$ or H_2SO_4 in alkylation reactions, and ionic liquids can be reused and recovered. Moreover, ionic liquids operate at substantially lower pressures and temperatures than conventional processes, saving money on both energy and capital expenditures. Ionic liquids have the capacity to function as catalysts and solvents.

4. Supercritical carbon dioxide ($scCO_2$)

When carbon dioxide is maintained at or above its critical pressure and critical temperature, it is said to be in supercritical condition. It is non-toxic, non-flammable, and inexpensive [26, 27]. Using hydrogen gas and supported palladium as the catalyst, heterogeneous catalytic hydrogenation of isophorene is one of the processes being carried out in $scCO_2$ (Scheme 17).

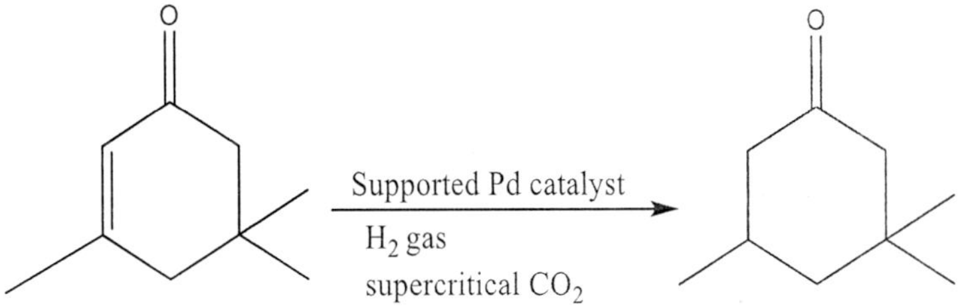

Scheme 17: Hydrogenation of isophorene in supercritical CO_2

In the microelectronics industry, $scCO_2$ is used instead of conventional organic solvents to spin-coat photoresists. Because of its low surface tension, $scCO_2$ is able to clean fine features on flat-panel displays and integrated circuits more effectively than traditional solvents. CO_2 can be compressed and recycled again. Supercritical CO_2 is used as a solvent for metal cleaning, dry cleaning, and textile processing. It replaces unsafe dry-cleaning chemicals such as perchloroethylene. It is used in the food sector to extract coffee's caffeine and red chili's colors. Once the pressure is decreased, the extracted compounds are recovered (Scheme 18).

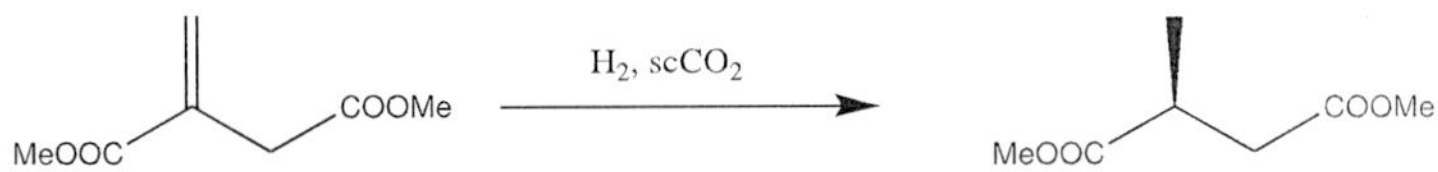

Scheme 18: Hydrogenation of alkene in supercritical CO_2

5. Supercritical Water (scH_2O)

The physical properties of supercritical water are radically different from those of ambient liquid water. The range of chemistry that can be performed in water is greatly increased by the characteristics of supercritical water (scH_2O), which include higher acidity, lower polarity, and lower density. ScH_2O is employed as a solvent in the formation of terephthalic acid. Conventional terephthalic acid synthesis uses acetic acid as a solvent with air and p-xylene at 190° C and 20 atm pressure with a manganese/cobalt catalyst system. Even though the procedure is very selective, it can be energy-consuming. In the process of the reaction, 10% of the acetic acid is oxidized as terephthalic acid is also insoluble in acetic acid. On the other hand, oxygen, terephthalic acid, and p-xylene are soluble in ScH_2O. The new technique uses a catalyst of 1000 ppm manganese dibromide and oxygen derived from hydrogen peroxide as the oxidizing agent. Terephthalic acid has an isolated yield of 70% and a selectivity of greater than 90% [28]. Table 1 shows green solvents.

Table 1: Green solvents

Sr. No.	Green Solvents	References
1.	Isoamyl alcohol	[29]
2.	2-Ethylhexanol	[30]
3.	2- Butanol	[31]
4.	Diethylene glycol butyl ether	[32]
5.	Ethyl acetate	[33]
6.	Dimethylpropylene urea	[34]
7.	Ethylene glycol	[35]
8.	t-Butyl acetate	[36]
9.	1-Butanol	[37]
10.	Propionic acid	[38]
11.	Butyl acetate	[39]
12.	n-Propyl acetate	[40]
13.	Methyl isobutyl ketone	[41]
14.	Isopropyl acetate	[42]
15.	Glycerol	[43]

Chemical technology for waste management

By using certain chemicals as oxidizers reducers, bases, acids, etc., waste is exposed to chemical treatment to lower its hazard. Hazardous wastes can sometimes be converted into useful by products through chemical treatment, which lowers the overall cost of waste disposal [44].

The techniques used in chemical technology include (Figure 3):

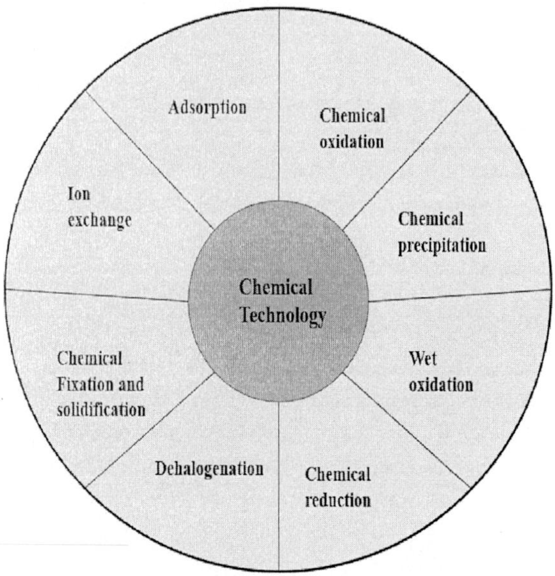

Fig 3: Chemical technology for waste management

1. **Adsorption:** The wastewater treatment technique of adsorption is used to separate various components from industrial effluent. In order to remove non-biodegradable organic compounds from groundwater, other water sources, process water, or drinking water preparation, adsorption is most frequently used as a tertiary cleaning step after biological water purification. For examples in order to remove active carbon adsorption, BTEX, PAH from groundwater, and chlorinated hydrocarbons, are widely utilized in soil remediation. For temporary (short-term) pumping or the purification of minute amounts, this technique is especially effective. Active carbon is used in the distribution of textiles to remove difficult-to-degrade substances like colorants.
2. **Chemical precipitation:** Chemical precipitation is a widely used and well-tested process for removing greases, suspended solids, oils, fats, and various other organic compounds (including organophosphates) from wastewater. In general, precipitation is the process of forcing dissolved or suspended impurities to form solid precipitates that can be filtered, spun down, or otherwise separated from the liquid component. During development and settling, a large precipitate can pick up ions and particles, effectively "sweeping" them out of the wastewater [45]. For example, impurities are removed from water using reagent coagulation, or chemical precipitation which precipitates them through coprecipitation electro-oxidizing potential, pH change, or utilizing precipitating agents (coagulants) such as ferrous or aluminum sulfates.
3. **Wet oxidation:** Organic molecules in wastewater can be treated using wet air oxidation (WAO). WAO is a destructive wastewater technique that oxidizes contaminants in the liquid phase at high temperatures and pressures. The organic and inorganic contaminants interact with the molecular oxygen dissolved in the wastewater [46].
4. **Chemical oxidation:** It is the process by which undesirable or harmful chemical species are transformed into chemical species that are neither undesirable nor harmful through oxidation (such as redox reactions) in wastewater treatment [47].
5. **Ion exchange:** The term "ion exchange" refers to the exchange of ions between two electrolytes or between an electrolyte solution and a compound. Ion exchange can be used for the separation, purification, and disinfection of aqueous or ion-containing

solutions. In wastewater treatment plants, ion exchange resins are employed to carry out demineralization by exchanging one ion for another [48].

6. **Chemical fixation and solidification:** In the stabilization and solidification (S/S) method of soil remediation, pollutants are made immobile by interactions with additives or other procedures. Contaminants may be chemically bound or encapsulated into a matrix during this procedure, also known as immobilization, fixation, or encapsulation [49].

7. **Dehalogenation:** The dehalogenation measurements could be performed concurrently or sequentially with the breakdown of plastic trash. It might be possible to de-halogenate waste plastics by co-grinding them with environment friendly additives like bio-wastes (like rice husk) and recyclable minerals (like red mud) via MCT for the production of solid fuels [50].

8. **Chemical reduction:** Waste is subjected to chemical therapy in order to reduce its hazard by using specific chemicals like oxidizers, acids, reducers, bases, etc. Chemical treatment enables the recovery of valuable byproducts from hazardous wastes, which lowers the total costs of waste disposal [51, 52].

Conclusion

Catalysts play a crucial role in green chemistry by reducing energy requirements, increasing selectivity, and enabling the use of less risky reaction conditions. It is indisputable that these catalysts are essential for controlling how a reaction proceeds and limiting or preventing the creation of side products.

We must concentrate on every element of the chemical reaction if we want to achieve nearly 100% "greenness" of chemical processes. Most corporate responses to environmental directives and protection up to this point have been of the circumstantial variety, like lowering human vulnerability to hazardous materials. Yet, an intrinsic technique (such as using the molecular design) can be not only more dependable but also less expensive. In this sense, some of the difficulties chemists face can be summed up as follows:

Use biodegradable, recyclable solvents and reagents, as well as reagents that are effective at catalytic concentrations. Choose solvents that can be quickly and readily separated later.

The publication of the hazardous waste rules has raised awareness about the management of hazardous wastes, however, the world has not yet implemented efficient procedures for safe management. The technical issues relating to waste treatment and disposal must be strengthened, and it must be determined what kind of technological input is needed, in order to achieve sustainable development.

References

1. Gladysz, J. A. (2001). Recoverable catalysts. Ultimate goals, criteria of evaluation, and the green chemistry interface. Pure and Applied Chemistry, 73(8), 1319-1324.

2. Clark, J. H. (2001). Catalysis for green chemistry. Pure and Applied Chemistry, 73(1), 103-111.

3. Anastas, P. T., Bartlett, L. B., Kirchhoff, M. M., & Williamson, T. C. (2000). The role of catalysis in the design, development, and implementation of green chemistry. Catalysis today, 55(1-2), 11-22.

4. Lancaster, M. (2020). Green chemistry: an introductory text. Royal society of chemistry.

5. Bubalo, M. C., Vidović, S., Redovniković, I. R., & Jokić, S. (2018). New perspective in extraction of plant biologically active compounds by green solvents. Food and Bioproducts Processing, 109, 52-73.

6. Olabi, A. G., Obaideen, K., Elsaid, K., Wilberforce, T., Sayed, E. T., Maghrabie, H. M., & Abdelkareem, M. A. (2022). Assessment of the pre-combustion carbon capture contribution into sustainable development goals SDGs using novel indicators. Renewable and Sustainable Energy Reviews, 153, 111710.

7. Burger, P., Plainfossé, H., Brochet, X., Chemat, F., & Fernandez, X. (2019). Extraction of natural fragrance ingredients: History overview and future trends. Chemistry & biodiversity, 16(10), e1900424.

8. Bubalo, M. C., Radošević, K., Redovniković, I. R., Halambek, J., & Srček, V. G. (2014). A brief overview of the potential environmental hazards of ionic liquids. Ecotoxicology and environmental safety, 99, 1-12.

9. Margallo, M., Taddei, M. B. M., Hernández-Pellón, A., Aldaco, R., & Irabien, A. (2015). Environmental sustainability assessment of the management of municipal solid waste incineration residues: a review of the current situation. Clean Technologies and Environmental Policy, 17, 1333-1353.

10. Misra, V., & Pandey, S. D. (2005). Hazardous waste, impact on health and environment for the development of better waste management strategies in future in India. Environment International, 31(3), 417-431.

11. Götz, M., Lefebvre, J., Mörs, F., Koch, A. M., Graf, F., Bajohr, S.,... & Kolb, T. (2016). Renewable Power-to-Gas: A technological and economic review. Renewable energy, 85, 1371-1390.

12. Sheldon, R. A. (2012). Fundamentals of green chemistry: efficiency in reaction design. Chemical Society Reviews, 41(4), 1437-1451.

13. Dobereiner, G. E., & Crabtree, R. H. (2010). Dehydrogenation as a substrate-activating strategy in homogeneous transition-metal catalysis. Chemical reviews, 110(2), 681-703.

14. Trost, B. M., Chapter 6 in Anastas, P. C. and Williamson, T. C., eds., Green Chemistry; Frontiers in Benign Chemical Synthesis and Processes. Oxford University Press, New York, 1998.

15. BHC Company, The Presidential Green Chemistry Challenge Awards program, summary of 1997 award entries and recipients, EPA 744-S97-001, U.S Environmental Protection Agency, Office of Pollution Prevention and Toxics, Washington, DC, p. 2, 1997.

16. Rotman, D. (1993). Mobil/Badger to market zeolite-based cumene technology. Chemical Week;(United States), 152(7).

17. Spindler, F., Pugin, B., Buser, H., Jalett, H. P., Pittelkow, U., & Blaser, H. U. (1998). Enantioselective catalysis for agrochemicals: synthetic routes to (S)-metolachlor, (R)-metalaxyl, and (αS, 3R) - clozylacon. Pesticide Science, 54(3), 302-304.

18. Rohart, E., Belliere-Baca, V., Yokota, K., Harlé, V., & Pitois, C. (2007). Rare earths-based oxides as alternative materials to Ba in NOx-trap catalysts. Topics in Catalysis.

19. Wan, Jie-Ping, Chunping Wang, Rihui Zhou, and Yunyun Liu. "Sustainable H 2 O/ethyl lactate system for ligand-free Suzuki–Miyaura reaction." RSC advances 2, no. 23 (2012): 8789-8792.

20. Breslow, Ronald. "Determining the geometries of transition states by use of antihydrophobic additives in water." Accounts of chemical research 37, no. 7 (2004): 471-478.

21. Breslow, Ronald, Kevin Groves, and M. Uljana Mayer. "Antihydrophobiccosolvent effects for alkylation reactions in water solution, particularly oxygen versus carbon alkylations of phenoxide ions." Journal of the American Chemical Society 124, no. 14 (2002): 3622-3635.

22. Biscoe, Mark R., and Ronald Breslow. "Hydrophobically directed selective reduction of ketones." Journal of the American Chemical Society 125, no. 42 (2003): 12718-12719.

23. Biscoe, Mark R., and Ronald Breslow. "Oxaziridinium salts as hydrophobic epoxidation reagents: remarkable hydrophobically-directed selectivity in olefin epoxidation." Journal of the American Chemical Society 127, no. 31 (2005): 10812-10813.

24. Welton, Thomas. "Room-temperature ionic liquids. Solvents for synthesis and catalysis." Chemical reviews 99, no. 8 (1999): 2071-2084.

25. Pavlinac, Jasminka, Marko Zupan, Kenneth K. Laali, and StojanStavber. "Halogenation of organic compounds in ionic liquids." Tetrahedron (Oxford. Print) 65, no. 29-30 (2009).

26. Sarbu, Traian, Thomas J. Styranec, and Eric J. Beckman. "Design and synthesis of low cost, sustainable CO2-philes." Industrial & engineering chemistry research 39, no. 12 (2000): 4678-4683.

27. Fukuoka, Shinsuke, Mamoru Kawamura, Kyosuke Komiya, Masahiro Tojo, Hiroshi Hachiya, Kazumi Hasegawa, MuneakiAminaka, Hirosige Okamoto, IsaburoFukawa, and Shigenori Konno. "A novel non-phosgene polycarbonate production process using by-product CO 2 as starting material." Green Chemistry 5, no. 5 (2003): 497-507.

28. Dunn, Jennifer B., Douglas I. Urquhart, and Phillip E. Savage. "Terephthlic acid synthesis in supercritical water." Advanced Synthesis & Catalysis 344, no. 3-4 (2002): 385-392.

29. Tran, T. T. V., Kongparakul, S., Karnjanakom, S., Reubroycharoen, P., Guan, G., Chanlek, N., & Samart, C. (2020). Selective production of green solvent (isoamyl acetate) from fusel oil using a sulfonic acid-functionalized KIT-6 catalyst. Molecular Catalysis, 484, 110724.

30. Kumar, V., Jahan, F., Kameswaran, K., Mahajan, R. V., & Saxena, R. K. (2014). Eco-friendly methodology for efficient synthesis and scale-up of 2-ethylhexyl-p-methoxycinnamate using Rhizopus oryzae lipase and its biological evaluation. Journal of Industrial Microbiology and Biotechnology, 41(6), 907-912.

31. Al-Shaal, M. G., Dzierbinski, A., & Palkovits, R. (2014). Solvent-free γ-valerolactone hydrogenation to 2-methyltetrahydrofuran catalysed by Ru/C: a reaction network analysis. Green Chemistry, 16(3), 1358-1364.

32. Liu, Y., Wang, Y., Zhai, C., Chen, W., & Qiao, C. (2014). Kinetics study of the esterification reaction of diethylene glycol monobutyl ether with acetic acid catalyzed by heteropolyanion-based ionic

liquids. Industrial & Engineering Chemistry Research, 53(38), 14633-14640.

33. Levere, M. E., Chambon, P., Rannard, S. P., & McDonald, T. O. (2017). MADIX polymerization of vinyl acetate using ethyl acetate as a green solvent; near-complete monomer conversion with molecular weight control. Journal of Polymer Science Part A: Polymer Chemistry, 55(15), 2427-2431.

34. Byrne, F. P., Jin, S., Paggiola, G., Petchey, T. H., Clark, J. H., Farmer, T. J.,... & Sherwood, J. (2016). Tools and techniques for solvent selection: green solvent selection guides. Sustainable Chemical Processes, 4, 1-24.

35. West, A. G., Barner-Kowollik, C., & Perrier, S. (2010). Poly (ethylene glycol) as a 'green solvent'for the RAFT polymerization of methyl methacrylate. Polymer, 51(17), 3836-3842.

36. Coomber, Charlotte E., et al. "Catalytic direct amidations in tert-butyl acetate using B (OCH 2 CF 3) 3." Organic & Biomolecular Chemistry 17.26 (2019): 6465-6469.

37. Folić, M., Gani, R., Jiménez-González, C., & Constable, D. J. (2008). Systematic selection of green solvents for organic reacting systems. Chinese Journal of Chemical Engineering, 16(3), 376-383.

38. Blohm, S., & Heinze, T. (2020). Mechanistic considerations of efficient esterification of starch with propionic anhydride/lauric acid in the green solvent imidazole. Macromolecular Chemistry and Physics, 221(23), 2000264.

39. Gao, F., Ferlin, F., Bai, R., Li, M., Vaccaro, L., & Gu, Y. (2021). Replacing halogenated solvents by a butyl acetate solution of bisphenol S in the transformations of indoles. Green Chemistry, 23(10), 3588-3594.

40. Byrne, F. P., Jin, S., Paggiola, G., Petchey, T. H., Clark, J. H., Farmer, T. J.,... & Sherwood, J. (2016). Tools and techniques for solvent selection: green solvent selection guides. Sustainable Chemical Processes, 4, 1-24.

41. Zha, J., Fan, B., He, J., He, Y. C., & Ma, C. (2022). Valorization of Biomass to Furfural by Chestnut Shell-based Solid Acid in Methyl Isobutyl Ketone–Water–Sodium Chloride System. Applied Biochemistry and Biotechnology, 194(5), 2021-2035.

42. Lawrenson, S., North, M., Peigneguy, F., & Routledge, A. (2017).

Greener solvents for solid-phase synthesis. Green Chemistry, 19(4), 952-962.

43. Wolfson, A., Dlugy, C., & Shotland, Y. (2007). Glycerol as a green solvent for high product yields and selectivities. Environmental Chemistry Letters, 5, 67-71.

44. Milligan, P. W. (1998). The anaerobic biodegradation of dicamba and related chlorosalicylates under different reducing conditions. Rutgers The State University of New Jersey, School of Graduate Studies.

45. EPA, U. (1980). Innovative and alternative technology assessment manual, EPA/430/9–78–009. US Environmental Protection Agency, Washington, DC.

46. Bhargava, S. K., Tardio, J., Prasad, J., Föger, K., Akolekar, D. B., & Grocott, S. C. (2006). Wet oxidation and catalytic wet oxidation. Industrial & engineering chemistry research, 45(4), 1221-1258.

47. Ksibi, M. (2006). Chemical oxidation with hydrogen peroxide for domestic wastewater treatment. Chemical Engineering Journal, 119(2-3), 161-165.

48. Kansara, N., Bhati, L., Narang, M., & Vaishnavi, R. (2016). Wastewater treatment by ion exchange method: a review of past and recent researches. ESAIJ (Environmental Science, An Indian Journal), 12(4), 143-150.

49. Razzell, W. E. (1990). Chemical fixation, solidification of hazardous waste. Waste Management & Research, 8(2), 105-111.

50. Shen, Y., Zhao, R., Wang, J., Chen, X., Ge, X., & Chen, M. (2016). Waste-to-energy: Dehalogenation of plastic-containing wastes. Waste management, 49, 287-303.

51. Misra, V., & Pandey, S. D. (2005). Hazardous waste, impact on health and environment for development of better waste management strategies in future in India. Environment international, 31(3), 417-431.

52. Rai, S., Wasewar, K., Mukhopadhyay, J., Yoo, C. K., & Uslu, H. (2012). Neutralization and utilization of red mud for its better waste management. World, 6, 5410.

Chapter - 4

Inherent Safety

Abstract

Since inherent safety was first introduced in the 1970s, there have been significant advancements in applying it to create safer chemical procedures, which are primarily characterized by strategic and concurrent improvement in environment, health, safety and economic performance. Hazardous or unforeseen events frequently occur during industries. Hazard should be seen as a flaw in design and actions ought to be taken to reduce or eradicate it during the designing process. Hazards can take many different shapes and have many different causes. Generally, this chapter provides a comprehensive image of inherent safety, which is anticipated to aid scientists and professionals specialising safety in chemical processes in comprehending the idea behind it and how effective it is at forming friendlier chemical processes. Making industrial processes as effective, clean, and secure as possible is a key goal of green chemistry in the creation of successful industrial ecosystems. The idea that green chemistry is naturally safe and clean has significant benefits for reducing environmental impact. Green chemistry avoids endangering both its practitioners and the ecosystem by utilising non-toxic substances and environmentally safe procedure.

Keywords: Inherent safety, Hazards, chemical process, economic performance

Introduction

The idea of inherent safety was first put forth in the late 1970s by Trevor Kletz. It is grounded in regular sense and involves refraining from using hazardous materials and engaging in hazardous inherent safety activities. Instead of generating conditions that will result in the threat, then attempting to manage it, an inherent safe procedure avoids it [1]. There are a variety of ways to accomplish this. Multiple risks can be connected to a chemical control procedure. Risks can occur as a result of properties of processes, the operation, complexity of the stages in the processes, and environmental circumstances, the expertise or degree of the staff's training, as well as raw supplies, intermediates, end products, by-products and waste materials [2, 3].

The chemical processes and interactions used for manufacturing chemicals are given top priority in green chemistry. Utilizing currently used chemical synthesis methods while improving their safety and pollution-reduction while also producing the necessary chemicals through greener methods is one way to make chemical synthesis more environment friendly [4, 5]. A simple instance of the former is switching out a volatile, toxic solvent for the reaction medium in a chemical synthesis process. In some circumstances, a reagent can be prepared more securely by using biological processes rather than chemical ones. Using various reagents for the synthesis that are safer and less likely to pollute is a second general strategy for making chemical preparations more environment friendly [6].

Hazards and its consequences

Any item, situation, or action that has the capacity to harm people, property and ecosystem is considered as a hazard [7].

The following consequences could result from a hazard:

1. **External fire:** possible triggers include the existence of pyrophoric material, flammable liquids, solids, metals, wood, or waste, as well as static friction, sparks, welding, lightning, hotspots, furnaces or auto ignition. Heat radiation, smoke, engulfment, fire damage and a cascading effect are the immediate effects.
2. **Internal fire:** A flammable gas, solid, liquid, metal, wood, or trash, pyrophoric material, halogen or oxygen, may be the origin of an ignition source such as hotspots, static, sparks, friction, lightning, welding, a furnace, or auto ignition. The likelihood of a flammable atmosphere forming in enclosed spaces is very high.
3. **Internal explosion:** may be brought on by an uncontrollable response, testing of the equipment, physical effort, cleansing, or filling. Instrument harm, Object/fragment damage and structural damage are the immediate results.
4. **Confined explosion:** may be brought on by an explosive or unstable substance, liquid, solid, flammable gas, dust, mist, oxygen or other of these. Equipment damage, missile/fragment damage, a cascading effect, and fumes/gases are the immediate consequences.
5. **Unconfined explosion:** like blast, explosion of gas or VCE (vapor cloud explosion) may be brought on by an explosive or unstable substance, flammable gas, liquid, solid, dust, mist, oxygen, or other of these. Immediate effects include a projectile, a commotion, illumination, a domino effect, and fumes or gases.

6. **Chronic exposure:** Acid, alkali, mist, liquid, gas or air may be the cause. Chronic effects on the public and workers are the immediate consequences.
7. **Pollution:** Gas escape, mist, acid, alkali, vapor, fume, algae, storage exhaust gas can all be the cause. Chronic impacts on workers and the general public as well as harm to flora and fauna are the immediate consequences.
8. **Violent energy release:** may be triggered by physical, potential or electrical energy. Equipment damage and a domino impact are the immediate consequences.
9. **Noise:** It may be brought on by ejectors, mechanical adjustments, vents, flares, alarms, pressure releases, or other devices. Noise that is louder than a certain decibel level is dangerous. Noise pollution and transient hearing loss are the immediate effects.
10. **Visual impact:** possibly brought on by an explosion or conflagration. The immediate result is transient blindness and inconvenience.
11. **Moving object:** road vehicles (tankers, articulated trucks), ramps, elevators, dumpers, forklift trucks, stackers, cranes, train wagons, or robots might act as triggers. Physical harm to employees or a breakdown of equipment are the immediate consequences.

Approaches for minimizing hazards

The four approaches for reducing the frequency of hazards are enumerated below (Figure 1) in decreasing order of robustness and dependability [8]:

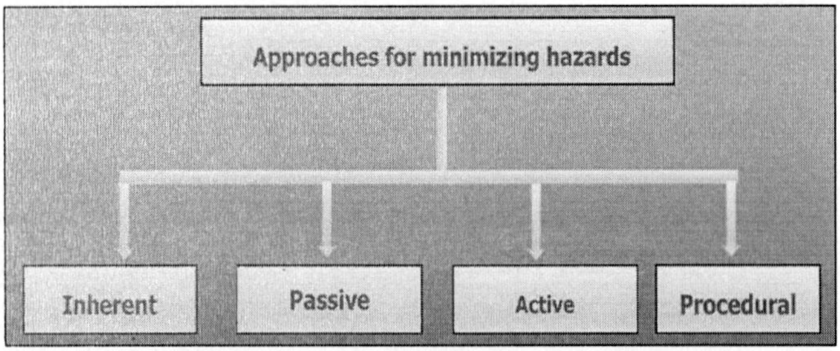

Fig 1: Approaches for minimizing hazards

1. **Inherent:** Employing less hazardous materials and production conditions to reduce or eliminate hazards. The inherent hazard reduction strategy is now supported because it is the most reliable way to provide safe working conditions.

2. **Passive:** Minimising risks by incorporating equipment and process design elements that reduced event frequency or outcome without the need for active device operation.
3. **Active:** To identify potentially dangerous process irregularities and to take quick remedial action, safety interlocks, process controls and emergency shutdown devices are used.
4. **Procedural:** To prevent accidents or mitigate their impacts, management approaches such as administrative controls, operating procedures, emergency response, and others are applied. When conducting complex processes like an atomic reactor, this is adhered very strictly.

Inherent safety principles

The principles of inherent safety are as shown in Figure 2 [9, 10]:

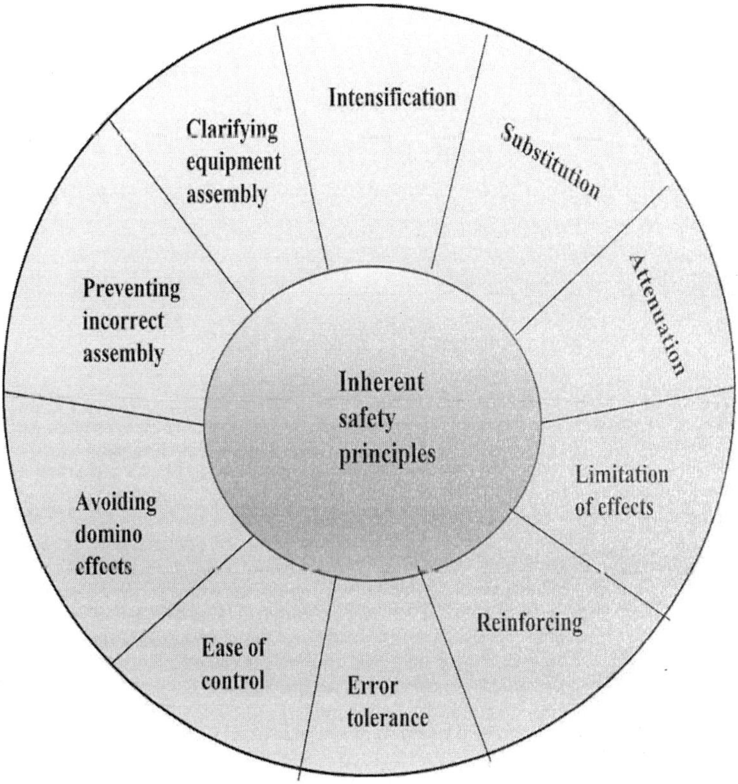

Fig 2: Inherent safety principles

1. Intensification

Involves reducing down on the amount of dangerous chemicals used in the plant or combining unit activities. For example, large quantities of the dangerous chemical no longer need to be stored or transported by building manufacturing plant which is alongside the ethylene oxide factory as the pipelines directly receive it [11]. This strategy reduced the severity of the potential risk. Combining unit projects operations reduces the amount of equipment needed while also lowering the risks. One of the primary sources of risk and danger in a chemical production process are reactors because they frequently operate with toxic and corrosive chemicals at high temperature and pressures. The reactor must also maintain the reaction mass throughout the reaction. Only with complete understanding of kinetics of reaction, mass transport, heat transfer and mixing, an optimal reactor design could be achieved. Despite the fact that the intrinsic reaction time is tiny, slow or insufficient mixing could lengthen it. Such processes necessitate cutting-edge reactor layouts, ideal mixing, and perhaps compact reactors. Chlorination was carried out in a batch stirred tank reactor and was replaced by a loop reactor with vigorous shaking and recirculation.

2. Substitution

Suggests the substitution of a less dangerous material for one that is more hazardous. It might be feasible to swap toxic chemicals for nontoxic ones and flammable solvents for non-flammable ones. Not only the substance's properties, but also the amount handled, must be evaluated. Instead of using a potentially dangerous chemical and adding numerous control alarms, safety checks and switching to a safer reagent might be a good option. This idea is illustrated by the substitution of air for hydrogen peroxide in oxidation processes and hydrogen for hydrazine in reduction reactions. Magnesium hydroxide slurry should be used instead of concentrated sodium hydroxide solution to control *pH*.

3. Attenuation

Implies lessening the difficulty of the operation. If it is necessary to use a dangerous chemical in a process, it must be determined whether the extremity can be mitigated by running the processes at a reduced pressure or temperature. It is possible to keep ammonia and chlorine as refrigerated liquids at atmospheric pressure as opposed to autoclaves, which operate at high pressure and ambient temperature. Both rate of leakage and vaporisation are reduced by the decreased pressure and temperature. Dilution is a different tactic that can result in attenuation; for example, using air as an oxidant rather than pure oxygen could stop explosive situations. The vapour cloud downstream can be reduced by 10 times when 28% aqueous ammonia solution is used instead of anhydrous ammonia in the event of a reactor leak.

4. Limitation of effects

Is accomplished by altering designing or processes conditions as opposed to introducing faulty protective equipment. For instance, it is preferable to use a lower-temperature fluid to prevent overheating rather than relying on a control mechanism to maintain overheating. When it comes to overheating, steam is less risky than oil because steam is constrained by its working pressure, whereas oil has the potential to overheat and may therefore need controllers to prevent it.

5. Containing/Enclosing/Reinforcing

Could be accomplished by separating workers, the general public, the environment's flora and fauna from potentially dangerous situations. To prevent leakage and human exposure, radioactive waste frequently gets encased with concrete and buried underground. Booms, which are movable barriers to oil, are typically used to contain oil spills in oceans and prevent them from contaminating the surrounding water. To prevent leakage, superfund sites are isolated from the surrounding area.

6. Error tolerance

Includes more durable machinery, procedures that can handle fluctuations in operating conditions such as temperature, pressure, and reactant concentration without triggering runaway reactions or exothermic conditions, or catalysts.

7. Avoiding domino effects

Domino effects are incidental or secondary effects that happen when an issue in one system spreads to other systems. A leaky valve in the storage vessel of a poisonous substance could contaminate other vessels downstream from it. The release of toxic fumes may result in immediate or long-term risks to people, plants and animals.

8. Preventing incorrect assembly

A lot of risks result from improper component assembly, which is caused by uncertainty in the designs of the subcomponents. The risk of human error during construction could be decreased or eliminated by using special assembly components. A frequent issue that causes flow to stop and pressure to build up is a non-return value that has been placed in the incorrect way.

9. Clarifying equipment assembly

Implies that it is crucial that others are fully aware of the situation, without any ambiguity, whenever a piece of machinery or a tool is opened up for servicing or repair, to prevent mistakes. Turning on the electricity while workers are still on the power line or turning on the utility flow before it has been finished or properly commissioned are two unsafe incidents that happen as a result of improper clarification of the equipment state.

10. Ease of control

Ease of control is correlated with less control and greater controllability.

Inherent Security

Following are the guidelines for inherent security [12]:

1. Perception

Reduce the amount of focus that plants attract. Tanks and vessels that are visible from the outside might make good targets.

2. Information

From a security perspective, it is better if there is less information accessible to the general public about the manufacturing site. The freedom to know must be balanced against the local communities' needs to know.

3. Plant layout

The delicate region has to be closely watched and kept unnoticed. Outsiders shouldn't be able to quickly identify control rooms.

4. Design

Large storage tanks and other vessels should be built with defence against aerial assaults.

5. Protection of the safety measures or devices

It is important to take the necessary precautions to prevent unauthorised individuals from simply disabling security systems.

6. Computer systems

The internet should not be linked to process control computers because doing so leaves them vulnerable to potential "cyber-attacks."

7. Buffer zone

Between a sensitive location with limited access and the general area that most plant employees use, a buffer zone needs to be established.

Industrial Examples

Green chemistry is a novel technique that is now under rapid development and offers us a proactive path for the sustainable advancement of science and technology in the future. When using the twelve green chemistry principles, chemical synthesis can be more effective and less harmful to the environment [13]. Figure 3 showing list of industrial examples.

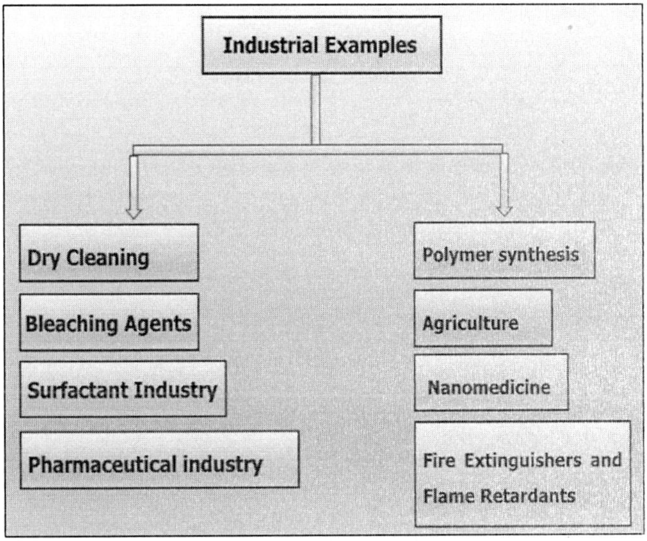

Fig 3: Industrial examples

i) Dry Cleaning

For dry cleaning, perchlorocthylene (PERC) is the most used solvent. It is known to be carcinogenic. In recent years, Micelltechnology [14], employs a surfactant and liquid CO_2 to dry clean clothing, thus eliminating the use of halogenated solvents [15].

ii) Bleaching Agents

The primary raw source utilised by paper and pulp industry is wood i.e., is made up of lignin and polysaccharides (70–30%). To produce high-quality paper, the wood's lignin content must be reduced. Different chemicals can used to breakdown lignin like Sodium sulphide, Sodium hydroxide and Chlorine gas, which breakdown 80–90% of the lignin but also pollute the environment and cause other issues. For the decomposition of lignin, chlorine reacts with lignin's aromatic rings to create dioxins including 2,3,4-tetrachloropdioxin and chlorinated furans. These substances are cancer-causing. Another more environment friendly approach uses hydrogen peroxide as the bleaching ingredient in place of chlorine, in the presence of various activators referred as Tetra amido macrocyclic ligands activators (TAMLs) [16]. Using this, H_2O_2 can degrade more lignin at low concentrations in less time [17].

iii) Surfactant Industry

The most often utilised products for both industrial and home usage are surfactants, soaps, and detergents. The majority of these products are non-biodegradable, and as a result, they remain in the water bodies.

Sodium silicate can take the place of phosphorus-containing chemicals in washing powder that are more environment friendly. For liquid cleaning applications, cationic and anionic surfactants can be abrasive; coconut oil soaps are better options. Rhamnolipid biosurfactants have been developed by the Jeneil Biosurfactant Co. in the USA. Rhamnolipids are organic glycolipids that *Pseudomonas aeruginosa*, a bacterium, produces in the soil and in plants. These rhamnolipid biosurfactants were replaced or be combined with man-made surfactants like nonyl-phenol ethoxylate, which have persistence in environment and toxicity. Surfactants based on sucrose are excellent emulsifiers and biodegradable. Utilising long chain fatty acid esters and sucrose, *C. antarctica* lipase is employed in transesterification processes at 80 °C. Chemical approaches cannot efficiently synthesize such surfactants without producing waste and a large number of degraded side products [18].

iv) Pharmaceutical industry

Green chemistry played important role in synthesis of pharmaceutical products such as Sildenafil Citrate, Ibuprofen, Sertraline, Riboflavin, Vitamin B3, 6-APA, L-DOPA, Paclitaxel and many more [19, 20].

Ibuprofen

The Boots Company (UK) manufactured and copyrighted the first ibuprofen industrial synthesis in 1960s. There are a lot of by-products and waste produced by this six-steps procedure. Due to atom effectiveness of about 40%, waste is produced. In the 1980s, starting with the same raw material as shown in scheme 1, BHC Company established a novel, three step, environment friendly industrial technique for the synthesis of Ibuprofen. This method has a high atom efficiency rate (77%) compared to earlier ones. Little waste and unwanted by-products are produced as a consequence [21].

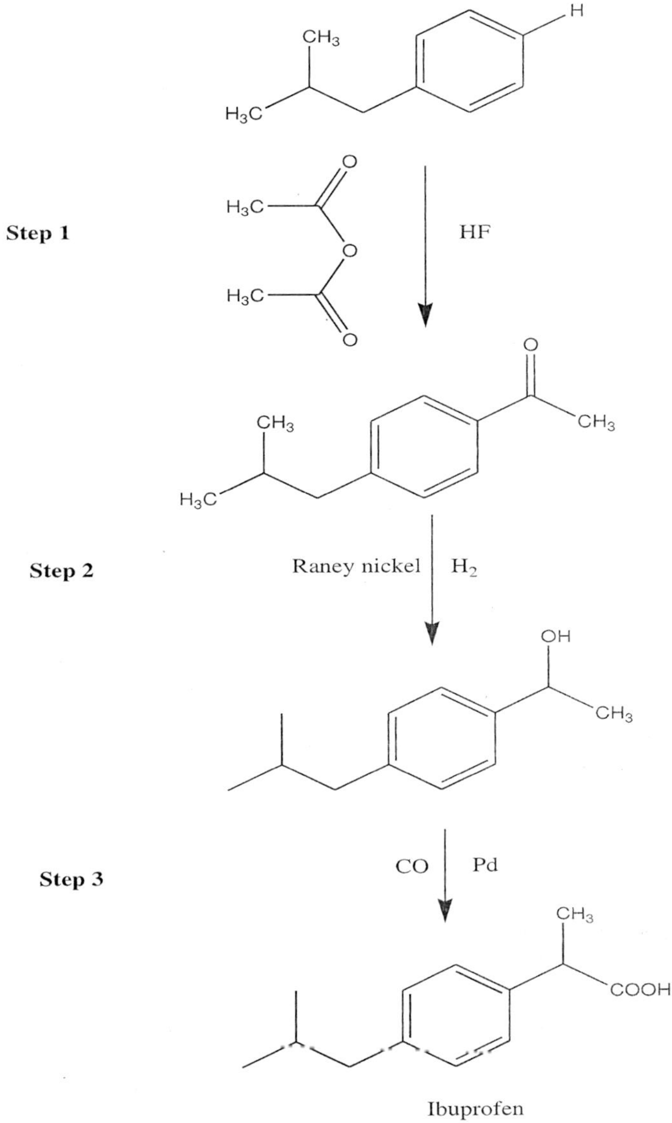

Scheme 1: Synthesis of Ibuprofen

Vitamin B3

The manufacturer of fine chemicals Lonza has developed a biotechnological pathway that starts with 3-cyanopyridine and ends with nicotinamide, better known as niacin or vitamin B3. Vitamin B3 synthesis involves green route, starting with 3-cyanopyridine to nicotinamide (also known as niacin or vitamin B3) shown in Scheme 2. The enzymatic hydrolysis of nitrile by *Rhodococcus* bacteria is the basis for conversions. The yields are quantitative, and the reactions are quite specific [22].

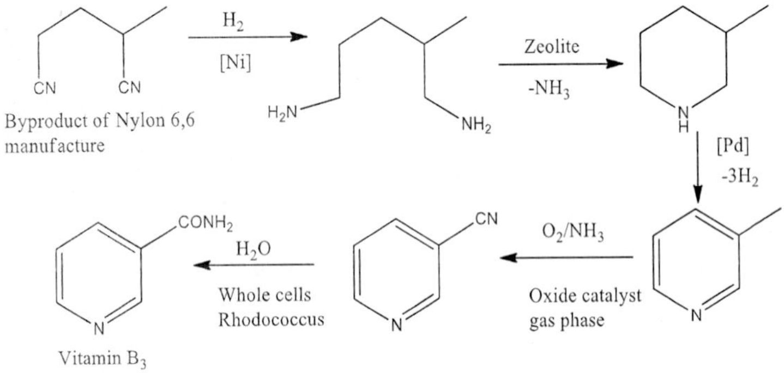

Scheme 2: Synthesis of Lonzanicotinamide

ACE Inhibitors

High blood pressure is treated with an ACE inhibitor known as captopril®. It is produced using the molecules L-proline and D-beta-hydroxy-isobutyric acid. Both of these building components are produced by fermentation, using the bacteria *Corynebacterium sp.* and the yeast *Candida rugosa* respectively. Then, using conventional chemistry, both raw ingredients are combined in the reactor to produce captopril as illustrated in Scheme 3. Milder conditions and less harmful chemicals are needed for the fermentation process [23].

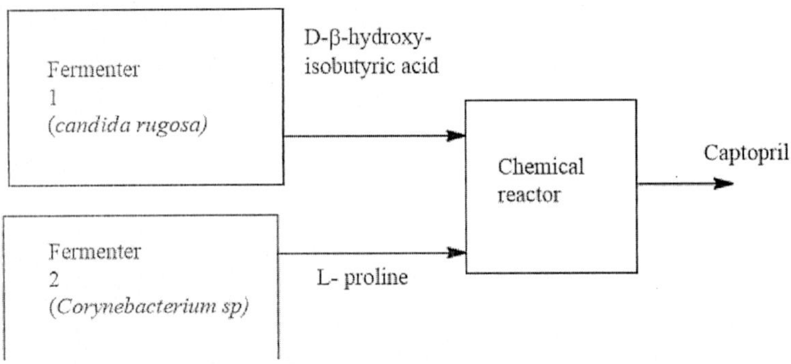

Scheme 3: Synthesis of captopril

v) Polymer synthesis

Traditionally, phosgene has been used to synthesize polyurethanes. Phosgene is a fatal and very poisonous gas. Without the use of phosgene, Monsanto had developed an alternative process for synthesising polyurethanes and their precursors (scheme 4). The procedure makes use of CO_2, a gas recognised to contribute to global warming (scheme 5). This procedure reduces the use of hazardous and poisonous chemicals while also reducing greenhouse gas emissions [24].

$$\underset{\text{(Amine)}}{RNH_2} + \underset{\text{(phosgene)}}{COCl_2} \longrightarrow \underset{\text{(isocynate)}}{RNCO} + 2\,HCl$$

$$RNCO \xrightarrow{R'OH} \underset{\text{(urethane)}}{RNHCOOR'}$$

Scheme 4: Synthesis of polyurethanes using phosgene

$$RNH_2 + CO_2 \longrightarrow RNCO + H_2O \xrightarrow[R'OH]{} RNHCOOR'$$

Scheme 5: Synthesis of polyurethane without using phosgene

The anionic polymer polyacrylic acid (PAC), which is used in many industrial processes, including the purification of water, it ends up in sewage since it is not biodegradable. Thermal polyaspartate (TPA) is a biodegradable, efficient, and economically viable substitute [25].

vi) Agriculture

Agriculture is one sector that can benefit from the application of green chemistry ideas. Eutrophication of water bodies has been caused by agricultural runoff, both pests and beneficial species are harmed by the use of broad-spectrum insecticides. A bio-pesticide called Bio-Cane® for sugarcane has been introduced in Australia. This product was created utilising a naturally occurring ring fungus that was grown on broken rice grains to act as a distribution medium. Bio cane granules have been marketed as being particularly effective against the greyback canegrub and as a replacement for conventional pesticides that harm the environment since they are poisonous, persistent in the soil, and leak into groundwater [26, 27].

One of the most extensively used herbicides is Roundup ®, produced by Monsanto (USA). Disodium iminodiacetate (DSIDA), a crucial step for its production, is often produced using Strecker's method. Hydrogen cyanide, formaldehyde, and ammonia are all needed for this procedure. Ammonia, formaldehyde, and hydrogen cyanide are poisonous substances that can potentially be harmful to the environment and human health. The new Monsanto technique, which employs copper catalyst to convert diethanolamine into DSIDA (Scheme 6), an intermediate of Roundup ®, has replaced the old one. This method uses no hydrogen cyanide, ammonia, or formaldehyde and is devoid of impurities and by-products, so it doesn't need to go through any additional purification procedures. Catalyst filtering allows the stream to be recycled [28].

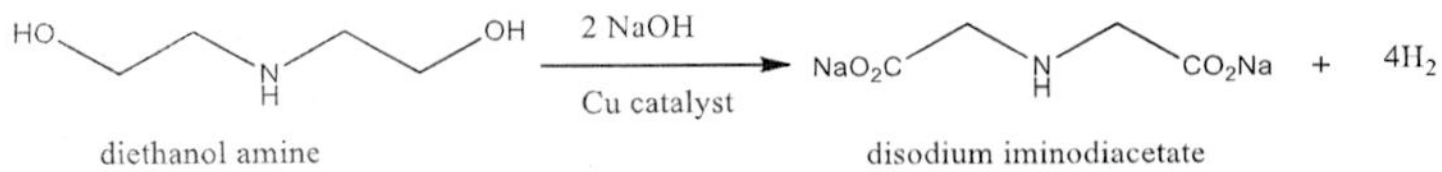

Scheme 6: Synthesis of DSIDA

vii) Fire Extinguishers and Flame Retardants

Previously, fire extinguishers were made with halogens (CFCs), well-known ozone-depleting materials. They taint water and destroy aquatic ecosystems. A non-toxic, biodegradable fire extinguishing and cooling chemical called Pyrocool can be used in place of conventional extinguishers. It does not destroy the ozone layer, extinguishes flames much more effectively than CFCs, and is not persistent in the environment [29].

viii) Nanomedicine

Green nanodrug delivery systems, which use natural biomaterials (such microorganisms and plant extracts) to create novel materials, are now transforming the nanomedicine industry [30, 31]. The current surge in improving human health is being driven by the use of nanoparticles (NPs) to deliver drugs to cells. These NPs are designed to only be drawn to damaged cells, allowing for direct therapy of those cells while increasing effectiveness, reducing side effects, and enhancing general human health. By using this method, pharmacological adverse effects are lessened in the body. Utilising plant extracts and biomaterials like lipids and proteins, the majority of green process innovations have produced materials with low toxicity and good biocompatibility [32].

Conclusion

Inherent safety means that the procedure is already secure and that no additional safety checks or alarms are required to keep it that way. Several hazard analysis methods are available and can be applied at different phases of the process development. Instead of adding layers of safety measures, it is preferable to design processes that are inherently secure. Industries constantly assess the costs and advantages of introducing safety components.

The first step in green chemistry is swapping out the chemical step for a biocatalytic step, which is then followed by totally swapping out the multi-step process for a single-step fermentation process. Additionally, researchers have also found that the processes used in biotechnology are generally mild, safe, and have minimal negative effects on the ecosystem. Given that the processes are orderly and selective, heterogeneous catalysis also seems to have a number of benefits.

References

1. Anastas ND, Warner JC. The incorporation of hazard reduction as a chemical design criterion in green chemistry. Chemical Health & Safety. 2005 Mar 1;12(2):9-13.

2. Dalzell G. Is operating cost a direct measure of inherent safety?. Process Safety and Environmental Protection. 2003 Nov 1;81(6):414-7.

3. Lo CC, Chen WJ. A hybrid information security risk assessment procedure considering interdependences between controls. Expert Systems with Applications. 2012 Jan 1;39(1):247-57.

4. Suardin J, Mannan MS, El-Halwagi M. The integration of Dow's fire and explosion index (F&EI) into process design and optimization to achieve inherently safer design. Journal of Loss Prevention in the Process Industries. 2007 Jan 1;20(1):79-90.

5. Edwards DW, Lawrence D. Assessing the inherent safety of chemical process routes: is there a relation between plant costs and inherent safety?. Process Safety and Environmental Protection. 1993 Nov;71:252-8.

6. Chen CY, Wu CW, Duh YS, Yu SW. An experimental study of worst case scenarios of nitric acid decomposition in a toluene nitration process. Process safety and environmental protection. 1998 Aug 1;76(3):211-6.

7. Edwards DW, Lawrence D. Assessing the inherent safety of chemical process routes: is there a relation between plant costs and inherent safety?. Process Safety and Environmental Protection. 1993 Nov;71:252-8.

8. Hendershot DC. Inherent safety strategies. Chemical Health & Safety. 1998 Jul 1;5(4):18-22.

9. Hendershot DC, Sussman JA, Winkler GE, Dill GL. Implementing inherently safer design in an existing plant. Process Safety Progress. 2006 Mar;25(1):52-7.

10. Hendershot DC. An overview of inherently safer design. Process safety progress. 2006 Jun;25(2):98-107.

11. ORRELL B, Cryan J. Getting rid of the hazard. CHEMICAL ENGINEER-LONDON. 1987 Aug 1(439):14-5.

12. Heikkilä AM, Hurme M, Järveläinen M. Safety considerations in process synthesis. Computers & chemical engineering. 1996 Jan 1;20:S115-20.

13. Anastas PT, Heine LG, Williamson TC. Green chemical syntheses and processes: introduction.

14. Blaser HU, Malan C, Pugin B, Spindler F, Steiner H, Studer M. Selective hydrogenation for fine chemicals: Recent trends and new developments. Advanced synthesis & catalysis. 2003 Jan;345(1-2):103-51.

15. Anastas PT, Williamson TC, editors. Green chemistry: frontiers in benign chemical syntheses and processes. Oxford University Press, USA; 1998.

16. Singh M. Future Outlook of Green Chemistry. International Journal of Green Chemistry. 2017;3(1):27-32.

17. Henkel M, Müller MM, Kügler JH, Lovaglio RB, Contiero J, Syldatk C, Hausmann R. Rhamnolipids as biosurfactants from renewable resources: concepts for next-generation rhamnolipid production. Process Biochemistry. 2012 Aug 1;47(8):1207-19.

18. Tundo P, Anastas PT, editors. Green chemistry: Challenging perspectives. Oxford University Press, USA; 2000.

19. Kirchhoff MM. Promoting sustainability through green chemistry. Resources, conservation and recycling. 2005 Jun 1;44(3):237-43.

20. Dunn PJ, Galvin S, Hettenbach K. The development of an environmentally benign synthesis of sildenafil citrate (Viagra™) and its assessment by Green Chemistry metrics. Green chemistry. 2004;6(1):43-8.

21. Cann MC, Connelly ME. The BHC Company Synthesis of Ibuprofen, a Greener Synthesis of Ibuprofen Which Creates Less Waste and Fewer Byproducts, chapter from "Real World Cases in Green Chemistry". Publisher American Chemical Society, Washington DC. 2000:19-24.

22. Straathof AJ, Panke S, Schmid A. The production of fine chemicals by biotransformations. Current opinion in biotechnology. 2002 Dec 1;13(6):548-56.

23. Hansson L, Lindholm LH, Niskanen L, Lanke J, Hedner T, Niklason A, Luomanmäki K, Dahlöf B, de Faire U, Mörlin C, Karlberg BE. Effect of angiotensin-converting-enzyme inhibition compared with conventional therapy on cardiovascular morbidity and mortality in hypertension: the Captopril Prevention Project (CAPPP) randomised trial. The Lancet. 1999 Feb 20;353(9153):611-6.

24. Anastas P, Hammond DG. Inherent safety at chemical sites: reducing vulnerability to accidents and terrorism through green chemistry. Elsevier; 2015 Nov 16.

25. Udayakumar GP, Muthusamy S, Selvaganesh B, Sivarajasekar N, Rambabu K, Sivamani S, Sivakumar N, Maran JP, Hosseini-Bandegharaei A. Ecofriendly biopolymers and composites: Preparation and their applications in water-treatment. Biotechnology Advances. 2021 Nov 15;52:107815.

26. Campos EV, Proença PL, Oliveira JL, Bakshi M, Abhilash PC, Fraceto LF. Use of botanical insecticides for sustainable agriculture: Future perspectives. Ecological Indicators. 2019 Oct 1;105:483-95.

27. Singh S, Sangwan S, Sharma P, Devi P, Moond M. Nanotechnology for sustainable agriculture: an emerging perspective. Journal of Nanoscience and Nanotechnology. 2021 Jun 1;21(6):3453-65.

28. Anastas PT, Bartlett LB, Kirchhoff MM, Williamson TC. The role of catalysis in the design, development, and implementation of green chemistry. Catalysis today. 2000 Jan 5;55(1-2):11-22.

29. Manley JB, Anastas PT, Cue Jr BW. Frontiers in Green Chemistry: meeting the grand challenges for sustainability in R&D and manufacturing. Journal of cleaner production. 2008 Apr 1;16(6):743-50.

30. Moond M, Singh S, Sangwan S, Devi P, Beniwal A, Rani J, Kumari A, Rani S. Biosynthesis of Silver Nanoparticles Utilizing Leaf Extract of Trigonella foenum-graecum L. for Catalytic Dyes Degradation and Colorimetric Sensing of Fe^{3+}/Hg^{2+}. Molecules. 2023 Jan;28(3):951.

31. Moond M, Singh S, Sangwan S, Rani S, Beniwal A, Rani J, Kumari A, Rani I, Devi P. Phytofabrication of Silver Nanoparticles Using Trigonella foenum-graceum L. Leaf and Evaluation of Its Antimicrobial and Antioxidant Activities. International Journal of Molecular Sciences. 2023 Feb 9;24(4):3480.

32. Jahangirian H, Lemraski EG, Webster TJ, Rafiee-Moghaddam R, Abdollahi Y. A review of drug delivery systems based on nanotechnology and green chemistry: green nanomedicine. International journal of nanomedicine. 2017;12:2957.

Chapter - 5

Future Trends in Green Chemistry

Abstract

Green chemistry is the process of creating compounds in a way that minimizes their resource and energy needs while producing little to no waste, making them safe, non-polluting, and environment friendly. Green chemistry is required to lessen the negative effects of anthropogenic materials and the processes used to make them. Green science is an appealing strategy for overcoming contamination aversion because it uses original logical solutions to circumstances that have happened in nature. Sustainable chemistry is another term for green chemistry. Sustainable chemistry is a method that encourages innovation across all industries to create and develop new chemicals, manufacturing processes, and sustainability management strategies that will increase performance and value while accomplishing the goals of safeguarding and increasing environment quality and human health. After a thorough review of the literature, ten areas of construction management that hold enormous promise for attaining sustainable growth have been selected and described in detail. These commandments will undoubtedly help to protect the ecosystem from severe damages.

Keywords: green chemistry, sustainable chemistry, sustainability, sustainable growth

Future aspects of green chemistry

Professor Ryoji Noyori, a Nobel Laureate, asserts that "Green Chemistry is not just a mere catch phrase; it is the key to the survival of mankind." Green chemistry has its roots in the late 1980s and early 1990s, when scientists and decision-makers first became aware of the detrimental effects that conventional chemical manufacturing processes had on the environment. As a result, a team of chemists led by John Warner and Paul Anastas started to develop the principles of "green chemistry" as a means of minimising the chemical industry's negative environment impact. Anastas and Warner's "Twelve Principles of Green Chemistry," which was released in 1998, provided an overview of the field's core concepts. A lot of research and development work has been done on green chemistry since the publication of the Twelve Principles, and it is now a recognised field of study. As a discipline, green chemistry is still developing and expanding today, with an increasing emphasis on sustainability and the creation of novel technologies and procedures that lessen the environment impact of the chemical industry [1].

According to green chemistry, industries should use environment friendly alternatives in place of any auxiliary chemicals, solvents or separation agents. To lessen waste in the processes, another step is to decrease reliance on fossil fuels and switch to renewable feedstock. Additionally, any negative environmental effects can be avoided by designing the chemicals and products so that they degrade into safe substances after use. Similarly, a single reaction can happen multiple times when catalysts are used rather than stoichiometric reagents. Lastly, real-time monitoring and control can be incorporated into the processes, which not only saves time but also prevents the formation of any harmful by-products. These factors make it safe to assume that green chemistry will play a substantial role in assisting chemists in creation of environment friendly and more effective production techniques. Industries must comprehend its advantages for their operations and adopt green chemistry to eliminate pollution, increase efficiency, conserve resources and achieve sustainable development goals.

The following are a few significant trends and advancements that will probably have an impact on green chemistry in the future:

1. **Increased attention to sustainability:** There will be a greater demand for sustainable and eco-friendly goods and procedures as people around the world become more aware of the need to protect the environment and natural resources. Green chemistry is in a good position to satisfy this demand because it is concentrated on the design and development of chemical materials and procedures that are sustainable and friendly to the environment [2].
2. **Continual technical advancement:** green chemistry will continue to profit from technological advancements, as new methods and tools are made available to support the creation of more environment friendly chemical goods and processes. This could involve the formation of novel materials, the utilisation of renewable resources and the use of data-driven methodologies and modern analytics to optimise chemical reactions [3].
3. **Growing regulatory support:** Governments all over the world are

starting to understand the significance of green chemistry and are acting to encourage the use of environment friendly and sustainable goods and procedures. This could involve establishing new laws and standards, providing incentives and subsidies, and establishing associations and working together with academia and industry [4].

4. **Increased collaboration and partnerships:** The success of green chemistry will depend on the ability of stakeholders from different industries to collaborate and share knowledge and expertise. This could involve partnerships among government, non-governmental organisations, and other stakeholders as well as collaborations between academia and industry [5].

Overall, green chemistry has a promising future as it develops and becomes more significant. As people around the world become more conscious of the necessity to safeguard the natural world and its resources, green chemistry is well positioned to play a substantial role in the development of environment friendly and sustainable processes and products that meet societal needs.

Green chemistry in sustainable development

In today's society, one of the most commonly used terms is sustainable development. "Sustainable development is growth that satisfies current needs of the present without compromising the capacity of future generations to satisfy their own needs." This statement applies to every aspect of society, but especially to chemistry and chemistry education. Both areas ought to consider how chemistry and chemistry education may promote greater sustainability in our society, both now and in the future. As a result, every action we make should be seen in the regard for social, economic, and ecological sustainability.

The promotion of a sustainable chemistry in research and industrial manufacturing is one of chemistry's contributions to address the challenge of greater sustainability in the growth of our society. Several efforts have been made to make future chemistry less toxic and dangerous under the banner of "green chemistry" (known as "sustainable chemistry" in Europe). Green chemistry attempts to improve chemistry's energy efficiency, decrease waste disposal, and produce novel products with minimal use of natural resources. New materials, products, and reaction routes are created, contributing to fulfil our needs today while also more carefully considering the needs of generations to come.

The United Nations released a global call to action in 2015 with the goal of eradicating poverty, protecting the environment, and ensuring that everyone lives in peace and prosperity by the year 2030. Governments, industry and many other international organisations are adopting this framework, which consists of 17 aspiring goals known as the Sustainable Development Goals (SDGs) in figure 1 [6]. Chemistry plays a vital role in achieving these Sustainable Development Goals. Possibilities for novel chemical investigation, green and sustainable chemistry instruction, green and sustainable chemical production techniques and a feeling of social accountability are essential for present and future chemists to safeguard our planet [7].

Fig 1: Seventeen sustainability goals

ACS (American Chemical Society) has designated seven priority SDGs that are essential to the activities of chemistry community.

Seven priority goals

1. Sustainable Development Goal 2: Zero Hunger

There are countless cases when chemistry can support humanity in achieving the aim of Zero Hunger. The improvement of food production and distribution processes, the extension of food shelf life through improved packaging, the preservation of food safety, quality and protection of plants from insects' manifestations are all possible by advancements in chemistry [8]. High-yield crops and novel fertilising techniques will enhance agricultural production and contribute to prevent soil erosion [9]. Food fortification contributes to the fight against malnutrition in communities where nutritious foods are scarce.

Some notable opportunities for the chemical industry to positively impact society include techniques for sustainable ammonia production, tailored crop protection agents, phosphate reuse and recovery, drought prevention for crops, and active packaging to stop food deterioration [10].

2. Sustainable Development Goal 3: Good Health & Well-Being

Chemistry is essential for achieving the goal of good health and wellness. Medical advancements and technology readily accessible by chemical science facilitates a deeper understanding of how illness and dangerous chemicals in our water, food, and environment disturb human health. Drug research and medical diagnosis depend heavily on chemistry, which helps people live healthier and longer lives [11].

Chemistry also offers novel approaches to reduce pollution and its negative effects on human health. Hazardous chemical pollution could be reduced or eliminated with the help of green and sustainable chemistry [12].

3. Sustainable Development Goal 6: Clean Water & Sanitation

Chemists will help society in a number of ways to achieve the goal of clean water and sanitary conditions. To ensure that everyone has access to clean and inexpensive drinking water, novel water filtration techniques and less expensive desalination procedures will be necessary [13]. Water quality is improved by using eco-friendly technologies and pollution-reduction strategies.

Metals and micro pollutants are removed using new low-energy, high-efficiency separation techniques. The chemical sector implements manufacturing procedures that reduce water consumption and waste disposal procedures that prevent pollution. Chemical technology can achieve this goal through processes like heavy metal removal and solar assisted desalination [14, 15].

4. Sustainable Development Goal 7: Affordable & Clean Energy

Chemistry assists to achieve the goal of affordable and clean energy by developing new materials for renewable energy, improving the energy efficiency of chemical processes, and developing cleaner fuel technologies. Advanced materials that are readily available on earth were developed by chemists for the production of renewable energy sources like photovoltaics, wind turbines, batteries, thermal energy collection, super-capacitors, and energy storage options. New separation techniques, improved process layout, and new catalysts are used in the chemical processing industry to increase manufacturing efficiency [16, 17]. Super-capacitors, sodium-ion batteries, lithium-sulphur batteries, and lithium-ion batteries are some examples of chemical technology for energy storage to meet this goal [18].

5. Sustainable Development Goal 9: Industries, Innovation & Infrastructure

To reach the objectives of almost all of the SDGs, chemistry research and innovation must be taken to a completely new level. The chemical science can significantly advance the industries, infrastructure, and innovation in at least three ways. First, the infrastructure can be advanced and production facilities can be redeveloped in the chemical industries to become more sustainable. Advanced materials and coatings are designed, synthesised, and produced by the chemistry community to increase the sustainability and reliability of infrastructure. Third, encouraging chemistry research that boosts innovation for commercial uses.

6. Sustainable Development Goal 12: Responsible Consumption & Production

The chemistry community has as one of its top priorities promoting responsible consumption and production. Chemical products help improve the efficiency and effectiveness of production processes across a range of industries, which improves water conservation and fuel efficiency efforts. Through improvements in waste management techniques, food packaging and additives, and the prevention of food loss, the chemical processing industry contributes to reducing the effects of consumption on the life cycle. By recycling and reusing molecules to complete the cycle in chemical manufacturing, the circular economy is made possible in downstream industries. New business models, products, and solutions that go beyond the idea of using a product only once are developed in order to achieve a stable circular economy. In order to do this, we must focus on issues related to energy consumption reduction and advancing the use of biobased and alternative feedstocks. Plastic recycling and the production of second-generation Bioethanol produced from residual woody biomass is an example of chemical technology used to achieve this goal [19, 20].

7. Sustainable Development Goal 13: Climate Action

Chemistry plays a role in achieving the sustainable development goal related to climate action. Advances in atmospheric chemistry are essential to our ability to forecast the severity and effects of climate change. Chemical research is essential for mitigating and adapting to climate change. By producing chemicals with low carbon emissions and helping other industries to reduce their own carbon output, the chemical industry is moving towards a low-carbon economy. By offering material and energy solutions, research and development efforts help the downstream customer transition to a circular and low carbon economy. Chemical industry helps another sector to become more resilient to risks associated with climate change [21]. Basic principles of green and sustainable chemistry to meet this goal are:

- Maximize resource effectiveness
- Eliminate & reduce hazards & pollution
- Designs systems thoroughly & utilising life cycle thinking

Ten commandments of sustainable development

Sustainability means living without compromising the planet's essential ecosystems. The main barriers to sustainability are (1) population growth that exceeds Earth's carrying capacity, (2) potentially destructive climate changes, (3) the availability of enough food, (4) the depletion of Earth's resources, (5) availability of adequate energy, and (6) contamination of the natural world with toxic and persistent substances. It won't be easy to get past these challenges and achieve sustainability, and there's no guarantee that humanity will succeed or even survive on earth in the long run. Adherence to a few key principles will be necessary to achieve sustainability. These can be condensed into the following ten sustainability commandments [21].

- Environmental sustainability must be considered in economics, governmental structures, religious beliefs, and individual life styles if human welfare is to be measured in terms of quality of life rather than merely the accumulation of material goods.
- Because the strain on a planet's support system is equal to the product of the number of people and the demand that each person places on the planet's resources, it is imperative to address both the population density and the demand that each place on the planet.
- Given that technology will be used in attempts to meet human needs despite the possibility of a global catastrophe, it is critical to recognise the anthrosphere as one of the five fundamental spheres of the environment and to create and maintain it with the goal of zero environmental impact and highest sustainability.
- Energy is a crucial component of sustainability, so it is essential to develop efficient, abundant energy sources with negligible to no environmental impact.
- The climate that supports life on Earth must be preserved, and solutions for inevitable climate change must be found.
- Considering each of the five environmental spheres, it is necessary

to maintain and enhance Earth's capacity for biological and food productivity.

- The demand for materials must be significantly reduced, and materials must be made from renewable resources, be recyclable and degradable if dumped in the environment.
- Toxic, hazardous, and persistent substances should be reduced in production and consumption, and should not be dumped into the environment; any waste dumped to disposal sites should be converted to non-hazardous forms.
- There are risks in not taking any risks, which must be acknowledged.
- It is imperative that all social classes and demographic groups receive education about sustainability; this education must also be disseminated through all media and be the responsibility of everyone with knowledge of sustainability.

1. **First Commandment: Environmental sustainability must be considered in economics, governmental structures, religious beliefs, and individual life styles if human welfare is to be measured in terms of quality of life rather than merely the accumulation of material goods.**

This commandment answers the crucial query, "What is happiness?" Nowadays, a lot of people determine their level of happiness based on their material possessions, such as their sport utility vehicle, house, and amount and quality of food. Sport utility vehicles consume an increasing amount of fuel from diminishing petroleum reserves, comfortable homes need more energy to keep them cool and warm, large lots divert limited farmland away from food production, having a home faraway from workplace requires lengthy journeys that take fuel and time, too many members of the present generation have consumed diet to the point of unhealthy obesity. With significantly less material and energy consumption that is currently the case in wealthier societies, good health, a balanced diet, physical comfort, fulfilling work, positive interpersonal relationships, and enjoyable cultural activities are all achievable. These are the things that really matter for happiness. Societies must recognise that happiness and wellbeing can be attained with significantly less material and energy consumption if sustainability is to be achieved.

Environmental economics examines issues related to sustainability, costs associated with pollution, resource economics, the advantages and disadvantages of pollution control, and the worth of ecological services. Individuals need to make an unexpected level of commitment to sustainability. Many people appear to have a voracious hunger for things and activities that use a lot of materials and energy. Individuals must incorporate sustainability into their belief systems in order for sustainability to be achieved. In fact, it would be very advantageous if protecting the environment and protecting Mother Earth's limited resources were to become a religion or be amalgamated into already-existing religions [21].

Governments plays a significant role in sustainability on all levels and across international borders. At the local level, sustainability-promoting laws and regulations are crucial. Local governments have set up recycling programmes for materials like paper, plastic, glass, and metals to lessen the amount of solid waste that needs to be disposed of. In many cases, only national governments have the authority and legitimacy to implement significant projects and promote the changes required for sustainability [22].

2. **Second Commandment: Because the strain on a planet's support system is equal to the product of the number of people and the demand that each person places on the planet's resources, it is imperative to address both the population density and the demand that each place on the planet.**

The equation used to express how much pressure is placed on planet's support system is:

$$\text{burden} = (\text{number of people}) \times (\text{demand for each person})$$

This equation demonstrates that in reducing human impact on the planet, it is necessary to consider both the population and the demand that each person places on the planet's resources. To achieve sustainability, both issues must be addressed. The first component of the equation can be controlled, but the second component, the demand per person, may be more challenging to control. Demand can be reasonably predicted by looking at the amount of carbon emitted per person each year, which represents fossil fuel consumption. Carbon emission per person is most in more industrially developed countries [22].

China and India, the two countries with the largest populations, have considerably lower carbon emissions per person. The demand for energy-intensive services and material goods will increases along with the economies of these two countries. For example, conventional economic systems would require a doubling of global petroleum consumption if China's citizens living standards were to equal the average of Mexico, a nation with a low standard of living. The carrying capacity of the Earth would be put under an impossible strain if the average Chinese person lived like the average American person. Naturally, ways must be found to satisfy the basic resource requirements per person in more developed countries. There must be ways to give residents of less developed countries a high quality of life without placing unwarranted demands on the planet's resources [23].

3. **Third Commandment: Given that technology will be used in attempts to meet human needs despite the possibility of a global catastrophe, it is critical to recognise the anthroposphere as one of the five fundamental spheres of the environment and to create and maintain it with the goal of zero environmental impact and highest sustainability.**

Some environment activists have a hostile attitude towards technology and technological solutions to environmental problems, which is one of the most contradictory viewpoints. There is no way that people will return to living in tepees and caves. Even though it is known that misusing technology can lead to disaster, it will still be used to try to meet human needs. Rejecting technology is both unrealistic and foolish. Technology will never go away.

Therefore, one of the challenges facing modern humanity is how to use technology without harming the environment or depleting the planet's resources. In doing so, it is necessary to acknowledge the anthrosphere, one of the five main spheres of the environment (Figure 2).

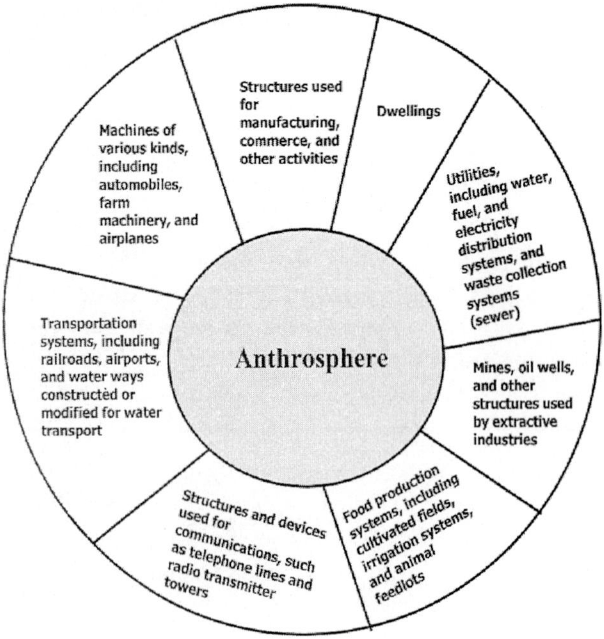

Fig 2: Anthrosphere: a complex system constructed and designed by humans

Reorientation of the anthrosphere is a necessity for sustainability so that it does not interfere with the sustainability and contribute towards sustainability. Both of these areas have a lot of potential for development. The design and operation of the anthrosphere is already well understood so that it does not interfere with sustainability. The anthrosphere must be run in a way that maximises material recycling, produces the minimum wastes possible, does not pollute the environment, and uses energy as effectively as possible. Additionally, to the greatest possible extent, energy and materials must be supplied from renewable sources.

The human environment can be planned and managed in a way that benefits and enhances other environment spheres. The majority of contemporary earth moving tools, for instance, have historically been used in ways that harm the environment, but they can also be used constructively to change the surface of the geosphere in ways that benefit the biosphere. Some particular steps that may be taken are the following [24]:

- Reconstruction of surface soil in regions where this resource has been depleted due to poor farming methods or contamination by waste and pollution.

- Building wetlands that can be used to clean wastewater to a standard that allows its discharge into the environment.
- Land terracing to prevent water-induced soil erosion.
- Water is pumped underground and refilled to replenish depleted aquifers.
- Building water storage impoundments.
- Sediment removal from bodies of water to re-establish aquatic life-friendly conditions.
- Building and running reverse osmosis plants to get rid of extra salt from irrigation water.
- Building electrified railroads to replace environmentally damaging, resource-intensive, and inefficient truck transportation.

4. **Fourth commandment: Energy is a crucial component of sustainability, so it is essential to develop efficient, abundant energy sources with negligible to no environmental impact.**

It is impossible to achieve sustainability without the creation of effective, long-lasting, and non-polluting energy sources. The biggest barrier to sustainability is found here because the primary energy sources currently in use, which are based on fossil fuels, are insufficient, unsustainable, and seriously pollute the environment. Fortunately, fossil fuel alternatives are available if there is a desire to develop them.

The sun ultimately fuels the majority of renewable energy sources. The most direct use of solar energy is solar heating. In recent years, solar heating of water and building has become more popular, and it should be used whenever possible. Using photovoltaic cells, it is possible to convert solar energy into electrical energy, and this practise is growing in popularity. Although the cost of solar energy is currently higher than that produced using fossil fuels, it is steadily declining and is already competitive in some remote areas without access to power grids.

Wind energy is now available as a somewhat unexpected alternative for fossil fuels and has become economically viable in many regions. The main drawback of solar and wind energy is that they are both intermittent. When the sun isn't shining, solar power is ineffective, and wind power is ineffective when the wind isn't blowing. To ensure an even flow of energy, it is essential to have reliable energy storage methods. Lead storage batteries are used to store energy in small installations, but they are impractical to be used for an entire power grid. It is also possible to store energy in the momentum of large flywheels spinning at very high speeds on almost frictionless air-lubricated bearings.

Moving water can be used to generate enormous amounts of energy. Worldwide, significant portions of total electricity are produced by hydroelectric installations. Since sunlight energy circulates water within the hydrologic cycle, using it to generate water power is actually another way to use solar energy. Even though conventional hydropower contributes significantly to the world's energy needs, it is likely that further development will have unfavourable environmental effects, particularly if dams are built on rivers that are currently flowing freely [25].

Because it provides constant power, nuclear energy does not cause the atmosphere to warm. It has the benefit of being an established technology that has produced significant quantity of electricity in some countries for many years. Nuclear reactors are used to power many submarines and ships in the U.S. Navy, and some of them have a 28-year lifespan. There has never been a reactor malfunction in the American fleet that put lives in danger or released a sizable amount of radioactivity. Modern nuclear reactors have passive safety features that can stop almost any incident that would release a significant amount of radioactivity, making a Chernobyl-style disaster impossible. Waste disposal is the most frequently mentioned problem with nuclear energy. At power plant locations, pools are used to store spent fuel rods from years of nuclear energy production. Since spent fuel activity declines most rapidly over time, this alternative is not as harmful as it may seem. The reprocessing of spent fuel made the nuclear energy a viable long-term energy source [26].

Automobiles, trucks, buses and ships that are not connected to a power grid need energy sources to operate. These vehicles require liquid fuels, which are currently petroleum-based, and all of which emit the greenhouse gas CO_2. The most environment friendly alternative for the production of liquid fuels is to make them out of biomass. Biomass is utilized in two ways for liquid fuels. The fermentation of sugar from sugarcane or sugar from grain starch results in the production of ethanol, which is one of them. Another choice is diesel fuel, which is made by esterifying plant oils, particularly soybean oil. These sources depend on a high-value raw material needed for food and can only be sustained only with significant government subsidies. The best alternative for converting biomass sources into liquid fuels is thermochemical gasification, produces a synthesis gas composed of CO and H_2. Different ratios of CO and H_2 can be combined to produce a variety of fuels, such as gasoline, methane, diesel fuel and jet fuel [27].

Some other alternatives for energy production, preservation, and utilization are summarized below:

- Generating electricity from water without the use of dams using techniques like turbines anchored in rivers.
- Utilisation of fossil fuels with underground CO_2 sequestration.
- Increased nuclear energy production using breeder reactors that produce more fissionable material than they consume.
- The replacement of diesel-powered locomotives with renewable electricity in rail systems.
- Use of methane (synthetic natural gas) produced from biomass as a source of fuel for automobiles.
- Energy from intermittent renewable sources can be stored using techniques like pumping hydrogen gas produced by electrolyzing water into underground storage sites.

5. **Fifth commandment: The climate that supports life on Earth must be preserved, and solutions for inevitable climate change must be found.**

Humans can harm the environment most easily and effectively by altering the atmosphere in a way that leads to widespread global warming. The most likely reason for this greenhouse effect is the release of CO_2 into the environment as a result of the burning of fossil fuels. There is convincing scientific evidence that human activities are surely increasing atmospheric CO_2 levels and causing global warming. These factors and the resulting climate changes poses a great challenge to human existence on Earth.

Fifth and the Fourth Commandments are closely related duc to the close connection between the increased CO_2 level and the use of fossil fuels and energy. Forest destruction reduces the capability of trees to fix carbon dioxide, and the decomposition of forest biomass residues adds more CO_2 to the atmosphere. Termites eating wood, anoxic bacteria growing in flooded rice paddies, and flatulent emissions of ruminant animals (cows, moose and sheep) all emit methane, a greenhouse gas. Some synthetic gases, particularly unbreakable fluorocarbons, are also potent greenhouse gases. The reduction of greenhouse gas-producing practices, particularly the burning of fossil fuels, is necessary to achieve sustainability [28].

Climate changes are inevitable if predictions of Earth's climate warming due to greenhouse gases are accurate. Therefore, adaptation to global warming and the resulting climate changes will be required. Some of the actions that must be taken are itemised below [21]:

- Shifting agronomic production from drought-stricken areas to those where climate change has improved growing conditions.
- Vast irrigation plans to combat the drought.
- Creation of heat- and drought-resistant crops.
- Evacuating people from low-lying coastal areas that are being inundated by rising sea levels brought on by ice melt and ocean warming.
- Building sea walls and other structures to offset the effects of rising sea levels.
- Increased use of saline water, such as growing algae in salt water rather than fresh water for the production of biomass energy.

6. **Sixth commandment: Considering each of the five environmental spheres, it is necessary to maintain and enhance Earth's capacity for biological and food productivity.**

Loss of biological productivity on earth has a negative impact on sustainability and could cause a massive human population starvation. Numerous human activities have been known to have a negative impact on biological productivity, however these effects have largely been concealed by incredible agricultural advancements like increased fertiliser use, the creation of extremely productive crop hybrids, and pervasive irrigation. The following are some of the factors affecting productivity [29]:

- Degradation of topsoil caused by harmful agricultural practises
- Large-scale paving of the land and urbanisation of the land
- Desertification in which formerly productive land is converted to desert
- Deforestation
- Air contamination that hinders plant growth

All environmental spheres must be considered in order to maintain and improve biological productivity. Topsoil must be protected in the geosphere; if it is lost, it will be nearly impossible to restore the ability of the land to produce biomass. It is possible to manage the hydrosphere in a way that boosts biological productivity. One method to increase the amount of water that seeps into groundwater aquifers and decrease the destructive rapid runoff of rainfall is to terracing the land.

It is important to maintain the climate in the atmosphere in a way that promotes high bio productivity by limiting global warming. The biosphere itself may be managed to increase biological productivity. With extremely productive crops, this has long been practised. On forested lands, high-yielding trees like some hybrid poplars can boost the production of wood and wood pulp. The fastest-growing tree species in northern temperate regions are hybrid poplars from the same genus as aspen or cottonwood trees. For some applications, they can even be harvested annually.

To maintain biological productivity, the anthrosphere must be managed properly. It is essential to avoid paving huge areas of useful land. In the anthrosphere, factories can produce fertilisers to boost biological productivity [24].

7. **Seventh commandment: The demand for materials must be significantly reduced, and materials must be made from renewable resources, be recyclable and degradable if dumped in the environment.**

Sustainable development requires decreased material demand. A lot is being done to lower the demand for materials, and even more reductions may be possible. An early 1900s image of a rail line shows poles supporting ten to twenty heavy copper wires, for carrying telegraph or telephone communication. Today, a single fibre optic strand of the size of a thread is capable of carrying a lot more data. A fingernail-sized circuit chip has replaced the heavy transformers and glowing vacuum tubes found in a 1948-vintage radio's circuitry. These examples signify dematerialisation and also depict material substitution. For instance, the conducting wires that can be replaced by fibre optic cables are made of scarce copper, while fibre optic cables are made of silica that is extracted from endless sand deposits.

Wherever feasible, renewable resources should be used to produce materials. This prefers wood as a material over plastics made from petroleum. Plastics and other materials can be made from wood and many other biomass sources. Natural rubber is preferable to petroleum-based man-made rubber from a sustainability perspective, and it is completely possible that genetic engineering advancements will make it possible to grow rubber-producing plants in regions where natural rubber production is not currently feasible.

Wherever possible, materials should be recyclable. The way that materials are used greatly influences their ability to be recycled. Metal recycling, for instance, becomes comparatively more challenging when plastics are used to bind metal components tightly. It is therefore beneficial to design machinery, such as electronic devices or automobiles, in a way that makes recycling easier [21, 30].

It is necessary to dispose of some materials in the environment. Household detergent is one such substance that gets dumped in wastewater. These substances should be easily biodegradable, typically through microbial action. Regarding degradability, detergents are a great example of a success story. After World War II, household detergents became widely used, but they contained a poor biodegradable surfactant called ABS (Alkylbenzene sulfonates) that caused massive foam beds to build up in sewage treatment facilities and receiving waters. To solve the issue of non-degradable surfactant in water, the ABS surfactant was changed to LAS (linear alkylbenzene sulfonate) surfactant, which is easily broken down by bacteria [31].

8. **Eighth commandment: Toxic, hazardous, and persistent substances should be reduced in production and consumption, and should not be dumped into the environment; any waste dumped to disposal sites should be converted to non-hazardous forms.**

The main principle of green chemistry is to prevent the production and use of toxic, persistent, and harmful substances as well as their release into the environment. Despite the fact that it is not always possible to completely avoid such harmful substances, significant progress has been made in the field of green chemistry. To reduce the amount of toxic and dangerous substances, extensive research is being done in the field of chemical synthesis. When such substances are required to be used because there are no other options, it is frequently possible to produce small amounts of the materials on demand, eliminating the need to keep large stocks of hazardous materials on hand.

Hazardous wastes that were disposed improperly have been a major cause of recent environmental issues. Hazardous waste should be disposed of in protected chemical landfills, according to modern practise. This tactic has two issues. The first is that landfills are not actually "secure" without exorbitant costs, and the second is that, unlike radioactive materials, which eventually degrade into non-radioactive substances, different resistant chemical wastes never completely degrade into non-hazardous substances. Installing monitoring infrastructure around facilities for disposing of hazardous waste and watching out for leaks and emissions is one part of the solution. However, issues could arise hundreds of years from now, leaving a bad legacy for future generations [32].

Therefore, before being disposed of, any wastes should first be transformed into non-hazardous forms. In order to achieve this, hazardous materials must be converted into forms that will not evaporate or leach into water, as well as organic materials. Co-firing hazardous wastes with fuel in cement kilns is a good strategy for achieving this goal because the organic materials are destroyed and the alkaline cement inactivates heavy metals and acid gas emissions [33].

9. Ninth commandment: There are risks in not taking any risks, which must be acknowledged.

Various things are dangerous by nature but there are no suitable alternatives. We must resist becoming so risk-adverse that we forbid engaging in activities that are dangerous but essential. Nuclear power is a good illustration. It is a very serious idea to use a "controlled atom bomb" to produce energy. However, given the alternatives of either continuing to burn a lot of fossil fuels, which would almost certainly cause climate change, or drastically reducing energy use, which would almost certainly lead to poverty and other negative effects, it is clear that going nuclear is the best course of action [34].

As a result, it's crucial to responsibly manage risk and use risky technologies. As mentioned above, nuclear power plants can be run safely with the right planning and management. Computer applications and modern technology can be effective tools for lowering risks. The use of computers in device and system designs can help engineers to anticipate risks and develop safer substitutes. Computerized control can allow processes, like those used in chemical production, to run safely. Computerized systems can be designed with redundancy to cover potential failures. Additionally, as computerised robotics develops, it becomes more and more feasible for expendable robots to perform risky tasks in hazardous environments where previously humans would have been required to take chances. Although avoiding risks is a laudable goal for sustainable chemistry and sustainable technology as a whole, it is important to remember that without being willing to take some risks, many valuable goals would never be accomplished. Early aviation would not have been possible without risk-takers, and today's commercial aviation systems would not be as generally secure and dependable. Numerous life-saving drugs would never reach the market without the risks associated with testing experimental medicines. Risks must be taken with caution, but avoiding risks altogether will result in a lack of progress in crucial areas that are required for sustainability [35].

10. **Tenth commandment: It is imperative that all social classes and demographic groups receive education about sustainability; this education must also be disseminated through all media and be the responsibility of everyone with knowledge of sustainability.**

Even though achieving sustainability is the primary challenge facing humanity, the majority of people have pathetically little knowledge of it. Education is crucial and important for achieving sustainability. Sustainability education needs to start early with primary school students and be incorporated into school curriculum.

The waste plastic, paper and aluminium cans that are collected from recycling bins in elementary schools have a small benefit, but the lessons about sustainability that these bins demonstrate are much more valuable. Every student receiving a university degree in chemistry should have some background in green chemistry, and every engineering graduate should have some familiarity with the fundamentals of green engineering. Educating people about sustainability's guiding principles in non-technical fields is highly significant. Sustainability education should be required for all professionals, including those in the legal, economic, political, and medical fields.

It is especially difficult to educate the general public about sustainability's principles and importance. The challenge of educating the general public about sustainability is greater because they have more options for information sources than a captive student body. The media and the internet play important roles in this regard. Unfortunately, compared to the considerable amount of time given in the media to the immoral antics of some fool, almost no time is given to sustainability, a topic that is extremely important to everyone's lives. Therefore, it is the responsibility of those with knowledge and interest in sustainability to spread the word via the internet and the media [36].

Conclusion

Green chemistry has the potential to significantly enhance planetary health as well as the welfare of communities and individuals. To create more sustainable and environment friendly products and processes, it is being used in a wide range of industries and sectors, including the pharmaceutical, cosmetic, agricultural, manufacturing, and energy industries. Green chemistry can significantly benefit society and the environment, but it also has some difficulties and restrictions. A group of novels, eco-friendly compounds that could replace dangerous chemicals is being developed by chemists. But it's not always simple. Finding alternatives for widely used chemical components like plasticizers, solvents, and polymers can take decades. It is possible to successfully balance the benefits and drawbacks of green chemistry and determine the best course of action for a particular circumstance by thoroughly considering its advantages and disadvantages. Despite a number of challenges, green chemistry has the potential to be beneficial.

References

1. Lenoir D, Schramm KW, Lalah JO. Green Chemistry: Some important forerunners and current issues. Sustainable Chemistry and Pharmacy. 2020; 18:100313.
2. Ganesh KN, Zhang D, Miller SJ, Rossen K, Chirik PJ, Kozlowski MC, Zimmerman JB, Brooks BW, Savage PE, Allen DT, Voutchkova-Kostal AM. Green chemistry: a framework for a sustainable future. Organic Process Research & Development. 2021;25(7):1455-9.
3. Zimmerman JB, Anastas PT, Erythropel HC, Leitner W. Designing for a green chemistry future. Science. 2020;367(6476):397-400.
4. Zuin VG, Eilks I, Elschami M, Kümmerer K. Education in green chemistry and in sustainable chemistry: perspectives towards sustainability. Green Chemistry. 2021;23(4):1594-608.
5. Kirchhoff MM. Promoting sustainability through green chemistry. Resources, conservation and recycling. 2005;44(3):237-43.
6. Thapa A. Sustainable Development Goal–6 (clean water and sanitation) Status and Challenges in Nepal. Journey for Sustainable Development and Peace Journal. 2023;1(1):21-35.
7. Axon S, James D. The UN Sustainable Development Goals: How can sustainable chemistry contribute? A view from the chemical industry. Current Opinion in Green and Sustainable Chemistry. 2018;13:140-5.
8. Galabada JK. Towards the Sustainable Development Goal of Zero Hunger: What Role Do Institutions Play?. Sustainability. 2022;14(8):4598.
9. Lambin A, Corpart JM. UN sustainable development goals and green chemistry, key points for sustainably innovating at Roquette, a global leader in plant-based ingredients. Current Opinion in Green and Sustainable Chemistry. 2018;13:137-9.
10. Arora NK, Mishra I. Current scenario and future directions for

sustainable development goal 2: A roadmap to zero hunger. Environmental Sustainability. 2022;5(2):129-33.

11. McCullough BP, Kellison T, Melton EN, editors. The Routledge handbook of sport and sustainable development. Routledge; 2022.

12. FallahShayan N, Mohabbati-Kalejahi N, Alavi S, Zahed MA. Sustainable development goals (SDGs) as a framework for corporate social responsibility (CSR). Sustainability. 2022;14(3):1222.

13. de Lázaro Torres ML, BorderiasUribeondo P, Morales Yago FJ. Citizen and educational initiatives to support Sustainable Development Goal 6: Clean water and sanitation for all. Sustainability. 2020;12(5):2073.

14. Qin H, Hu T, Zhai Y, Lu N, Aliyeva J. The improved methods of heavy metals removal by biosorbents: A review. Environmental Pollution. 2020;258:113777.

15. Li C, Goswami Y, Stefanakos E. Solar assisted sea water desalination: A review. Renewable and Sustainable Energy Reviews. 2013;19:136-63.

16. Křížková A. Sustainable Development Goal 7: Affordable and Clean Energy Panel Data Analysis.

17. Zhou G, Xu L, Hu G, Mai L, Cui Y. Nanowires for electrochemical energy storage. Chemical reviews. 2019;119(20):11042-109.

18. Chen TL, Kim H, Pan SY, Tseng PC, Lin YP, Chiang PC. Implementation of green chemistry principles in circular economy system towards sustainable development goals: Challenges and perspectives. Science of the Total Environment. 2020;716:136998.

19. Requejo A, Peleteiro S, Rodríguez A, Garrote G, Parajó JC. Second-generation bioethanol from residual woody biomass. Energy & Fuels. 2011;25(10):4803-10.

20. Karpudewan M, Ismail Z, Roth WM. Ensuring sustainability of tomorrow through green chemistry integrated with sustainable development concepts (SDCs). Chemistry Education Research and Practice. 2012;13(2):120-7.

21. Manahan SE. Green chemistry and the ten commandments of sustainability. 2006.

22. Wilkinson A, Hill M, Gollan P. The sustainability debate. International Journal of Operations and Production Management. 2001;21(12):1492-1502.

23. Ritchie H, Roser M, Rosado P. CO_2 and greenhouse gas emissions. Our world in data. 2020.

24. Manahan SE. Environmental chemistry. CRC press; 2017.

25. Omer AM. Energy, environment and sustainable development. Renewable and sustainable energy reviews. 2008;12(9):2265-300.

26. Holechek JL, Geli HM, Sawalhah MN, Valdez R. A global assessment: can renewable energy replace fossil fuels by 2050?. Sustainability. 2022;14(8):4792.

27. Vasudevan P, Sharma S, Kumar A. Liquid fuel from biomass: an overview. 2005.

28. Suh S. Are services better for climate change?. 2006.

29. Manahan SE. Fundamentals of environmental chemistry. CRC press; 2011.

30. Worrell E, Carreon JR. Energy demand for materials in an international context. Philosophical Transactions of the Royal Society A: Mathematical, Physical and Engineering Sciences. 2017;375(2095):20160377.

31. Lin SH, Leu HG. Operating characteristics and kinetic studies of surfactant wastewater treatment by Fenton oxidation. Water Research. 1999;33(7):1735-41.

32. Ozbay G, Jones M, Gadde M, Isah S, Attarwala T. Design and operation of effective landfills with minimal effects on the environment and human health. Journal of environmental and public health. 2021;2021.

33. Ferronato N, Torretta V. Waste mismanagement in developing countries: A review of global issues. International journal of environmental research and public health. 2019;16(6):1060.

34. Kurniawan TA, Othman MH, Singh D, Avtar R, Hwang GH, Setiadi T, Lo WH. Technological solutions for long-term storage of partially used nuclear waste: A critical review. Annals of Nuclear Energy. 2022;166:108736.

35. Bates DW, Gawande AA. Improving safety with information technology. New England journal of medicine. 2003;348(25):2526-34.

36. Devi MN, Rao MN. Green Chemistry–Challenges and Opportunities Mrs. N. Gayatri Devi Mr. NVNB Srinivasa Rao. Chief Patron.:86.